SCIENCE, EVOLUTION AND CREATIONISM

EVOLUTION STANDS FAITH UP

REFLECTIONS ON EVOLUTION'S WARS

SCIENCE, EVOLUTION AND CREATIONISM

Additional books in this series can be found on Nova's website under the Series tab.

Additional e-books in this series can be found on Nova's website under the e-book tab.

SCIENCE, EVOLUTION AND CREATIONISM

EVOLUTION STANDS FAITH UP

REFLECTIONS ON EVOLUTION'S WARS

GUILLERMO PAZ-Y-MIÑO-C, PH.D.

NOVINKA

New York

For permission to use material from this book please contact us:
Telephone 631-231-7269; Fax 631-231-8175
Web Site: http://www.novapublishers.com

Library of Congress Cataloging-in-Publication Data

ISBN: 978-1-62948-440-2

Published by Nova Science Publishers, Inc. † New York

CONTENTS

PREFACE

> "...Shot-gun marriages between evolution and faith have never worked, despite the tradition of pointing the barrel at evolution's head. In truth, evolution likes it single. Free, with no stoppers of thought or restrains on logic. And when lured unknowingly into the altar by those who see facts and fiction compatible, evolution has consistently stood belief up and walked away, sometimes run, toward its secular turf... [The] dream of arranging evolution's wedding with belief will remain dormant for as long as evolution is awake."

I wrote this fact in 2011 after submitting myself to the torture of reading the last of a trilogy of spiritual books authored or sponsored by Francis Collins, Director of the National Institutes of Health NIH and former head of the Human Genome Project: *The Language of God* 2006, *Belief: Readings on the Reason for Faith* 2010, and the brand new —at that time— and coauthored with fellow evangelical Christian Karl Giberson, *The Language of Science and Faith*.

I posted a review of the 2011 "monology" in Amazon.com and rated it "one star" (of five possible stars a reviewer can award to a meaningful volume). Mine was one of two low ratings, matched by two "two stars" and outnumbered by three "three stars," seven "four stars," and fourteen "five stars."

What still surprises me is that up to August 2013 —as I write this Preface— fifty two readers have assessed my review, surpassing the top most read "five stars" rating, which has accumulated thirty nine reads (thirty four of them considering it "helpful"). However, only one in five bookworms —nowadays "cyberworms"— actually liked my review which, to me, was illuminating. It implied that twenty percent of all readers —since more read my posting than others'— did agree with the proposal that evolution and "belief" are incompatible. And that was encouraging.

In all the research on patterns of acceptance of evolution in the United States, Europe and Latin America (2009-2013), conducted in collaboration with Dr. Avelina Espinosa, a molecular microbiologist at Roger Williams University, we have found ubiquitous support to the "incompatibility hypothesis" (IH), which proposes that the controversy over evolution-and-science versus creationism is inherent to the incompatibility between scientific rationalism/empiricism and the belief in supernatural causation. Religiosity disrupts, delays and/or stops the correct comprehension and acceptance of evidence; it is *the* chief variable negatively correlated with acceptance of evolution and understanding of science. When faith in a deity is absent, like among atheists, non-believers and agnostics, acceptance of evolution is 100%. [Well, all people should accept the concept of evolution!].

But creationist books, like the trilogy above, were not the major justification for compiling, in this volume, my recent essays about evolution and its wars against superstition, which have been published scatterly, between 2010 and 2013, in *The Standard Times* of New Bedford, the *Evolution Literacy* website, *The Providence Journal*, and *Secular World* magazine. It was witnessing, as a university professor, educator, public speaker and editorialist, the pervasive decline in science education in my much-admired United States, the potential collapse of America's academia, and my frustration in attempting to reverse, at least around my immediate circle of influence, such trend.

I came to the US in 1992, as a graduate student, after obtaining a bachelors degree in Biology at the Pontifical Catholic University of Ecuador [yes, I am one of those *suma cum laude* atheists educated at an internationally prominent Jesuit institution] and a Diploma in Endangered Species Management at a joint program between the Jersey Wildlife Preservation Trust and the University of Kent at Canterbury, in the UK. In America, I immediately became fascinated with the university-system, the rigor in the conceptual debate about science and sophistication in the experimental work, the high standards —still fashionable in the early 1990s— in college education, the respect for curiosity-driven research, the critical thinking inside and outside the classrooms, and the passion for exploring the truth just because we wanted to seek it. Graduate school and postdoctoral training are [will always be] my most treasured educational experiences.

But parallel to this universe of intellectual opportunities in the US, I also witnessed the consolidation of Intelligent Design during the mid 1990s, a doctrine born in the 1980s, which had proposed that a Designer was responsible, ultimately, for the assemblage of complexity in biological systems; according to ID, evolution could not explain holistically the origin of the natural world, nor the emergence of intricate molecular pathways essential to life, nor the immense phylogenetic differentiation of biological diversity, and instead ID proposed an *intelligent agent* as the ultimate cause of nature. In conceptually mistaken, type-I-error-based arguments to discredit evolution, ID had attributed randomness to molecular change, deleterious nature to single-gene mutations, insufficient geological time or population size for molecular improvements to occur, and invoked "design intervention" to account for complexity in molecular structures and biological processes.

To my relief, I saw, in 2005, the demise of ID in court, during the Dover, Pennsylvania, trial (Kitzmiller et al. versus Dover School District, see link below) for violating the rules of science by "*invoking and permitting supernatural causation*" in matters of evolution, and for "*failing to gain acceptance in the scientific community*." A significant victory for evolution in its historical wars against "belief," although more symbolic than real. [Legislation challenges to evolution continue].

During the process of ripping ID apart, earlier variants of creationism resuscitated —mostly in media-driven discussions, which I never considered harmless since they reflected the quiescent mind of the public— and new emerged as, allegedly, better alternatives to ID. Among the former were Theistic Evolution and Creation Science, creationism in principle and practice (i.e. God, the maker of the universe, always present in the fore- or background of causality); among the latter was BioLogos (mid 2000s), which aimed at merging Christianity with science by proposing a "*model for divinely guided evolution*" that "*[required] no intrusions from outside for its account of God's creative process, except for the origin of the natural laws guiding the process.*" Supporters of BioLogos "*[suggested] that once life arose, evolution and natural selection permitted the development of biological diversity and complexity*," including humans. After evolution got under way, "*no special supernatural intervention was required*" (quotes from *The Language of Science and Faith*, above). In essence, God was done, but remained in touch for eternity!

That is why, in 2011, I stated that "...shot-gun marriages, like this, between evolution and faith, have never worked, despite the tradition of pointing the barrel at evolution's head..." and that "...evolution likes it single." But a more insightful statement was, perhaps, that "in matters of

God's non-existence, the high-school educated atheist is more lucid than the deeply religious scientist." And this will remain as truth for eternity.

Additional motivation to compile, in this book, my essays on evolution's wars came from seeing around my students and the public a mosaic of nonsense pseudoscience available "open access" for their consumption. From faith healing and rejection of medical advances in vaccination or birth control (Essay 2), to forecasting Armageddon, first for May 21, 2011, to later —after an "adjustment in dates"— October 21 of the same year (Essay 3). Or from the stubborn position of denying our ape, African ancestry by Young-Earth Creationists or simply by the misinformed public (Essay 14), to documenting high religiosity among the New England faculty and educators of prospective teachers at thirty five of the elite colleges and universities in Northeastern US, whose beliefs in "the supernatural" distorted their understanding of reality and likely polluted their teaching practices (Essay 17). Or from the 25th world-placement in mathematics of our students, 17th in science, 14th in reading (Essay 16), to the decline in science standards among nineteen states in the US (from their otherwise letter grades of B or A to lower scores) and the unchanged twenty states, half of them remaining around F or D by 2012 (Essay 18). Or from the 2012 declaration, by a US congressman, member of the House Committee on Science, Space and Technology, that "*God's word is true. I've come to understand that. All that stuff I was taught about evolution and embryology and the big bang theory, all that is lies straight from the pit of Hell*" (Essay 20), to the emergence of antievolution legislation as in the 2012 New Hampshire bill proposals to "*require evolution to be taught as a theory, including the theorists' political and ideological view points and their position on the concept of atheism*" or "*require science teachers to instruct pupils that proper scientific inquire results from not committing to any one theory or hypothesis, no matter how firmly it appears to be established.*" Both bills were [still are] hymns to ignorance (Essay 18).

The assemblage of this book is also inspired in my passion for biology, its grandeur and richness in history (Essay 11), its atomic connections with the beginings of the cosmos (Essays 7 and 8), its life forms in unique, unforgettable environments, like the Galapagos and Hawaiian archipelagos (Essays 4, 5 and 6); and with my addiction to the splendid American museums (antidotes to their impostor rivals "Genesis parks") where nature and the universe are explained to anyone willing to learn about the truth, like the Boston Museum of Science and its Charles Hayden Planetarium (Essay 7), or the romantic New Bedford Whaling Museum (Essay 10), where evolution is not only reassured to its vistors by displaying gigantic whales' skeletons, but also by connecting them to human history, to the "pain and suffering of nature," embodied by the slaughtering of whales, to secure prosperity of the young American republic (1820-1860) and sustain its Industrial Revolution. An ecocide that, although must not be forgotten, "lit the world" by fueling with cetacean oil, wax and baleen the economy of a nation that, during the 1900s, became a superpower.

Science is just a refined device for resolving ordinary curiosity and a powerlful liberator of superstition. It stands alone in its secular turf.

Guillermo Paz-y-Miño-C

Essay 1

EVOLUTION STANDS FAITH UP: ON FRANCIS COLLINS' & KARL GIBERSON'S "THE LANGUAGE OF SCIENCE AND FAITH"*

"…the dream of arranging evolution's wedding with belief will remain dormant for as long as evolution is awake."

In matters of God's nonexistence, the high-school-educated atheist is more lucid than the deeply religious scientist.

A trilogy of spiritual books, inspired in personal revelation, has been written or sponsored by Francis Collins, Director of the National Institutes of Health NIH and former head of the Human Genome Project: *The Language of God* 2006, *Belief: Readings on the Reason for Faith* 2010, and the brand new —coauthored with fellow evangelical Christian Karl Giberson— *The Language of Science and Faith* 2011.

Much "language" for a redundant message: "*The fool says in his heart, 'There is no God,'" PSALM 14:1*, reads the heading script to the fifth chapter of Collins' —hopefully— last "consilience" volume; a sort of "the pot calling the kettle black" or the blind accusing the sighted of not seeing the fullness of the empty glass. And empty indeed of evidence, except for that which the blind imagines.

But Collins and Giberson allege to navigate brightness and shuffle the pages of the book of nature, epitomized by Charles Darwin's Origin of Species 1859, with those of the Christian Bible, and propose a "*model for divinely*

---

* Essay first published in *Evolution Literacy*, 2011; also posted in *Amazon.com* as a book review.

guided evolution" that "requires no intrusions from outside for its account of God's creative process, except for the origin of the natural laws guiding the process." And the authors "*suggest that once life arose, evolution and natural selection permitted the development of biological diversity and complexity,*" including humans.

After evolution got under way, Collins and Giberson conclude, "*no special supernatural intervention was required.*" God was done, but remained in touch for eternity!

Shot-gun marriages like this, between evolution and faith, have never worked, despite the tradition of pointing the barrel at evolution's head. In truth, evolution likes it single. Free, with no stoppers of thought or restrains on logic. And when lured unknowingly into the altar by those who see facts and fiction compatible, evolution has consistently stood belief up and walked away, sometimes run, toward its secular turf.

C. S. Lewis' *Mere Christianity*, 1952, impacted Francis Collins, and deeply, while Collins was searching, during the late 1970s, to "*believe in some sort of God*" and to abandon his non-theist past, as he confesses in *The Language of God.*

Impossible to be fully comprehended, Lewis' God is the supreme architect of existence. God is both nature and its master. And Lewis' narrative can be persuasive, severe, and even frightening. *The Screwtape Letters* 1942, *The Great Divorce* 1946, and *The Problem of Pain* 1962, reveal a complex, pre and post World War II traditionalist and suffered Irish survivor who, as a teenager, used to be "*very angry with God for not existing.*"

But Lewis yielded to his inner struggles. In *An Experiment in Criticism* 1961, a book on how to judge literature and artistic expression, he wrestled with himself: "*The first demand any work of art makes of us is surrender. Look. Listen. Receive. Get yourself out of the way. There is no good asking first whether the work before you deserves such a surrender, for until you have surrendered you cannot possibly find out.*"

Francis Collins also "surrendered" to Jesus Christ on "*a beautiful fall day," as he "was hiking in the Cascade Mountains*" during his "*first trip west of the Mississippi,*" where "*the majesty and beauty of God's creation overwhelmed*" his "resistance" to accept a deity. As he "*rounded a corner and saw a beautiful and unexpected frozen waterfall, hundreds of feet high,*" he "*knew the search was over.*" The next morning, he "*knelt in the dewy grass as the sun rose*" and became a Christian.

The *Homo* centric perception of the cosmos —in the past byproduct of ignorance, at present of faith— has led Collins to infer "fine tuning in the universe to embrace life" —a principle traceable to Islamic philosophy of the eight and fifteen centuries [see also Victor J. Stenger book *The Fallacy of Fine-Tuning*, resources below]— but this too is illusory. It is actually life, its biochemical processes, its proximate and ultimate causation that are finely tuned to the effects of cosmic change.

Can a scientist really be a believer? In *Decoding The Language of God* 2010, George Cunningham responds, both as comrade physician and geneticist, to Collins' theistic evolution, repackaged as *BioLogos*, or the exploration, promotion and celebration of the integration of science and Christianity. Contradictions bleed out of Collins' book as Cunningham scalpels —in reality he butchers— the fallacies inherent to finding logic where there is nonsense.

Faith aside, Collins' outstanding research career and administrative skills shine. NIH and the Human Genome Project will always treasure his golden fingerprints of astute visionary policy. But his dream of arranging evolution's wedding with belief will remain dormant for as long as evolution is awake.

SUGGESTED READINGS AND RESOURCES

Collins, F. S. (2010). *The language of life: DNA and the Revolution in Personalized Medicine*. New York, NY: Harper Collins Publishers. [Contrary to *The Language of God* (2006), this "antithesis" book is sober and valuable].

Cunningham, G. C. (2009). *Decoding the language of god: Can a scientist really be a believer?* New York, NY: Prometheus Books.

Martin, M. (2006). *The improbability of god*. Amherst, NY: Prometheus Books.

Paz-y-Miño-C., G. & Espinosa A. (in review). The incompatibility hypothesis: evolution vs. supernatural causation. In: G. Trueba et al. (Eds). *Why does Evolution Matter? The Importance of Understanding Evolution*. Cambride, UK: Cambridge Scholars Publishing.

Paz-y-Miño-C., G. (2012). Evolution stands faith up. *Secular World* 18, 38.

Scott, E. C. (2009). *Evolution vs. creationism: An introduction*. Westport, MA: Greenwood Press.

Sharlet J. (2008). *The family: The secret fundamentalism at the heart of American power*. New York, NY: HarperCollins Publishers.

Stenger, V. J. (2012). *God and the folly of faith: The incompatibility of science and religion*. New York, NY: Prometheus Books.

Stenger, V. J. (2011). *The fallacy of fine-tuning: Why the universe is not designed for us.* New York, NY: Prometheus Books.

Essay 2

FAITH HEALING VS. MEDICAL SCIENCE*

> "…At times of nanotechnology medicine, where single-atom-thick sheets of carbon (graphene) can be injected into a body with the mission of cauterizing cancerous breast tissue if stimulated by laser, or when vaccination can immunize an entire country and prevent human papillomavirus from spreading among sexually active teenagers, or when evolutionary principles enlighten our understanding of disease and cure, pray healing cannot replace nor supplement, in urban or rural communities, scientific medicine."

"If a sense of disease produces suffering, and a sense of ease antidotes it, disease is mental. Hence the fact in Christian Science that the human mind alone suffers, and the divine mind alone heals it."

This metaphor belongs to Mary Baker Eddy (1821-1910), author of *Science and Health with Key to the Scriptures* (1875) and founder in 1879 of the Church of Christ, Scientist, whose modern, cathedral-like headquarters, the Mother Church —completed in 1904, emerged in downtown Boston rivaling in architecture the Trinity Church, the Holy Cross Cathedral, and even the Massachusetts Statehouse.

During a three-year search for the "divine laws of life" within the Bible, Eddy compiled passages about healing and envisioned a cure method based on prayer, which by the end of the 1880s propagated among the students at her Massachusetts Metaphysical College, a short-lasting enterprise (1881-1889). Her charismatic personality seeded the Christian Science Journal (1883), the Christian Science Sentinel (1898), the Herald of Christian Science (1903), and

* Essay first published in *The Standard Times* of New Bedford, September 2, 2010.

the Christian Science Monitor (1908), currently a dynamic online news survivor of historic financial struggles.

Like most spiritual therapies, Eddy's "pray healing" faded, although the enormous Mother Church of Christ Scientist she inspired still breathes in the heart of Boston through a majestic golden pipe organ, which gives the impression of resounding even in silence.

In fact, deity-mediated physical and mental well-being are inconsistent with modern medicine. However, the positive effects of the "relaxation response" —a mind-and-body state of calmness which is elicited during meditation and monotonous behavior like chanting or bead-praying— on the recovery from depression, anxiety, insomnia or pain seem biologically rooted and scientifically measurable.

The Institute for Mind Body Medicine at the Massachusetts General Hospital has sponsored the most compelling research on the relaxation response (RR). Cardiologist Herbert Benson described RR in 1974 as changes in metabolism, heart rate, respiration, blood pressure and brain chemistry triggered during the meditative state.

RR counters the fight-or-flight response, conceptualized by Harvard physiologist Walter Cannon in 1915 as an animal's ability to cope with danger via, not surprisingly, opposite mechanisms to RR. In essence, the fight-or-flight response excites stress response, while the relaxation response calms, thus bringing the organism to homeostasis.

Benson and his collaborators ignited three decades of investigations on RR which included the physiological changes that occur during RR, the cognitive-behavioral and psychological variables associated with it, and the diversity of meditation methods. Most recently, gene expression induced by RR has been the focus of Benson's exploration. In all scenarios, Benson and his colleagues have been cautious to insert RR as another variable into the complex equation of mind-body health.

Indeed, good research often demystifies popular certainty. In a 2006 study of the therapeutic effects of intercessory prayer in 1,800 cardiac bypass patients, sponsored by the Templeton Foundation —famous for surpassing the Noble Prize by granting $1.5 million to explorers of life's spiritual dimensions— Benson and his team reported a higher incidence of complications among patients who knew Christian devotees were praying for their recovery, in contrast to patients uncertain about receiving such prayers, who convalesced.

The data published in the *American Heart Journal* disappointed supporters of proximal or distant pray-mediated healing.

The Templeton Foundation, however, further committed $150,000 to study the effects of prayer on auditory and visual impairments in rural Mozambique, and religious studies professor Candy Brown, from Indiana University, embraced the task. In the September 2010 issue of the *Southern Medical Journal*, Brown and collaborators report 24 cases of improvement in hearing and/or vision after intercessory prayers.

The authors themselves confess the flaws of the study: unknown source of the impairments, unconfirmed diagnosis of ear or eye malfunction, patients' cultural habituation to healers, no control group, and a sample size 75 times smaller than that of Benson's. In sum, much enthusiasm and poor science, but the authors go on to state that prayer "*may be a useful adjunct to standard medical care... in contexts where access to conventional treatment is limited.*"

Benson's research on RR has brought into scientific scrutiny the belief of pray healing and provided a rational explanation for the sense of joy, well-being and calmness induced by meditation and its equivalents.

At times of nanotechnology medicine, where single-atom-thick sheets of carbon (graphene) can be injected into a body with the mission of cauterizing cancerous breast tissue if stimulated by laser (Nano Letters, 2010), or when vaccination can immunize an entire country and prevent human papillomavirus (HPV) from spreading among sexually active teenagers, or when evolutionary principles enlighten our understanding of disease and cure, pray healing cannot replace nor supplement, in urban or rural communities, scientific medicine.

Suggested Readings and Resources

Benson, H., Dusek, J. A., Sherwood, J. B., Lam, P., Bethea, C. F., Carpenter, W., Levitsky, S., Hill, P. C., Clem, D. W., Jain, M. K., Drumel, D., Kopecky, S. L., Mueller, P. S., Marek, D., Rollins, S. & Hibberd, P. L. (2006). Study of the Therapeutic Effects of Intercessory Prayer (STEP) in cardiac bypass patients: a multicenter randomized trial of uncertainty and certainty of receiving intercessory prayer. *American Heart Journal*, 151, 934-942.

Institute for Mind Body Medicine at the Massachusetts General Hospital. http://www.massgeneral.org/bhi/.

Wells, S. (2010). *Pandora's seed: The unforeseen cost of civilization*. New York, NY: Random House.

The Christian Science Monitor. http://www.csmonitor.com/.

Essay 3

WRONG AT FORECASTING ARMAGEDDON*

> "...It is the human mind through which the universe understands and predicts itself. And it is the scientific method, which civilizations have discovered and perfected over millennia, the most reliable approach to understanding the past and present, and hypothesizing about the future. There will never be Judgment Day or Rapture or Armageddon. Our planet shall remain cosmically unsafe with fluctuating danger until our sun, the cause of our existence, runs out of fuel within the next 5 billion years."

Scripture is a silly source of insight to forecast the end of the world. But science helps us hypothesize more fascinating phenomena than the delusional Rapture.

If human extinction comes abruptly, it shall be from the darkness of our galaxy or from within the Earth. In fact, cosmic collisions and mega eruptions have decimated life more than once during Earth's 4.6 billion-year existence. Ubiquitous nuclear explosions and the effects of radiation could also wipe out people and most organisms, but I will not elaborate here on such threats, since they have been examined by scholars and speculated in the media.

We can predict the probability of occurrence of some events based on prior empirical observations. I am certain that, for example, all readers of this article will not be around in the year 2100 [I have post-script doubt of this assertion since new data suggests a few of you, the readers, may make it to happy one-hundreds]. I can forecast with less confidence that 20% of my college students' grandparents, ages 60-70, will pass away within this decade, and that one in every five students' excuses for missing an exam —due to a "death in the family"— will be legitimate. And I can infer from survivorship

---

* Essay first published in *The Standard Times* of New Bedford, June 3, 2011.

tables compiled by the U.S. Centers for Disease Control and Prevention that Judgment-Day spiritualist Harold Camping, age 89 —who just postponed his "The End" of the world from May 21 to October 21 of this year [2011]— is reaching his own finale.

Redundant natural phenomena have led human intellect to discover predictable patterns of reality and, in consequence, reject superstition. And this is true about the cycles of the moon, the emergence and diversification of life from simple to more complex forms, the continental migrations of birds, the blooms of magicicadas every 13-17 years, or the seasonal intensification of tornadoes and hurricanes. Science has awarded us the luxury of anticipating the maladies of the blood (hemophilia) or the mind (porphyria) among the inbred Royal Houses of Europe, as much as foreseeing the geographic journey of the next influenza outbreak.

The self-correcting character of science guided astronomers to debunk Ptolemy's geocentric model of the years 100 —when the Earth was worshipped as "the heart" of the universe— and replaced such optical illusion with Copernican heliocentrism —16th century— an obvious deduction for those who refused to imagine heavens and constellations in the sky and rather saw planets and their moons revolving around our closest star, the sun.

Astronomy and modern cosmology have made sense of the chaotic night sky and allowed us to broadcast the passage of comets, alignment of planets, meteorite falls and polar auroras triggered by the collision of solar-wind particles against the atmosphere.

But why should we care about predicting cosmic events? Because by witnessing their majestic beauty and immensity we can also study and assess their destructive potential; and an educated public in matters of real global survival is more precious and needed than a vociferous street crowd surrendering at the brainless certitude of Camping's prediction of Armageddon.

The NASA Eclipse Website offers an impressive education playground for those curious about previous eclipses of our sun and moon, plus accurate schedules of upcoming eclipses up to 2015 [updates reach year 3000]. And NASA's Asteroid and Comet Impact Hazards Program informs online surfers about the predictability and reasonable uncertainty concerning future collisions between Earth and cosmic debris.

Watching the sky for possible encounters with large asteroids is a priority considering our planet's heritage: A 6-mile asteroid cremated the "ruling class" Reptilia 65 million years ago when blasting a 110-mile-wide crater in the Yucatan Peninsula of Mexico. Few reptile lineages survived and repopulated.

The end of the dinosaurs dragged tiny mammals out of their burrows, and they revisited old reptilian professions with limited originality: they became arboreal climbers, terrestrial herbivores and hunters, cave flyers, aquatic divers, and killers of killers. Their major evolutionary innovations were sweat glands turned into milk-producing udders and breasts, internal eggs that matured within a womb, and large skulls encasing clever brains. And it was the primate brain, shaped even further by natural selection in ape descendants, us, the thinking apes, that evolved scientific reasoning and consciousness.

It is the human brain through which the universe understands and predicts itself. And it is the scientific method, which civilizations have discovered and perfected over millennia, the most reliable approach to understanding the past and present, and hypothesizing about the future. There will never be Judgment Day or Rapture or Armageddon. Our planet shall remain cosmically unsafe with fluctuating danger until our sun, the cause of our existence, runs out of fuel within the next 5 billion years.

Suggested Readings and Resources

Centers for Disease Control and Prevention, data and statistics by topic. http://www.cdc.gov/DataStatistics/.

NASA, Asteroid and Comet Impact Hazards Program. http://impact.arc.nasa.gov/index.cfm.

NASA, Eclipse Web Site. http://eclipse.gsfc.nasa.gov/eclipse.html.

National Science Foundation, revisiting Chicxulub, a broad look at the evidence for a dinosaur-killing impact.
http://www.nsf.gov/news/news_summ.jsp?cntn_id=116480.

Rushton, A. R. (2008). *Royal Maladies: Inherited diseases in the royal houses of Europe*. Victoria, BC: Trafford Publishing.

Essay 4

UNFORGETTABLE GALAPAGOS, A SUMMIT, AND WHY EVOLUTION MATTERS*

"...There is no place on Earth like the Galapagos Islands and no better destination to discuss the reality of evolution. Charles Darwin visited this volcanic archipelago in 1835; the rest is 180 years of science's history after a major paradigm shift from creationism —including all its modern forms: theistic evolution, creation science, young-earth creationism, Intelligent Design, BioLogos— to evolution."

Evolution is true regardless of our awareness of it. The evolutionary processes preceded the emergence of our ape consciousness —which is only about 150,000 years old. Evolution itself is indifferent to our level of understanding of it, although we must admit that proper education leads to accepting evolution and admiring one of the most magnificent phenomena ever discovered, that of organisms' transformations by means of natural selection as described by Darwin in *The Origin of Species* (1859).

"Why Does Evolution Matter" was the theme of the III World Evolution Summit which just adjourned in San Cristobal Island, Galapagos, after a fascinating scientific gathering (June 1-5, 2013) organized by University San Francisco of Quito (USFQ) and its Galapagos Institute for the Arts and Sciences (GAIAS), Ecuador. Twelve international keynote speakers rationalized over evolution at the microscopic and organismic levels, the applications of evolutionary principles for the conservation of endangered species and their environments and, of course, for the relevance of evolution to human health.

---

* Essay first published in *Evolution Literacy*, June 9, 2013.

USFQ and GAIAS shinned big: they led us to 'dream' about evolution... Yes, one can dream about reality, feel the joy of understanding the facts about life's origins, from the tinniest molecules like RNA (an evolutionary precursor of DNA) to intriguing viruses who rely on RNA or DNA to infest each other or replicate inside cells and thus fuel unfolding biological complexity... and to the emergence of bacteria and archaea (prokaryotes, cells without a nucleus) and of nucleated unicellular organisms (eukaryotes), like amoeba, which reside freely in ponds or in the guts of reptiles and other creatures... and to the magnificent Galapagos hawks, penguins inhabiting the cold waters of this equatorial archipelago –away from their ancestral home in the South Pole— and sea lions resting on the shores occasionally awakening to hikers' traffic. Indeed, uniquely amazing, beautiful, unforgettable.

The keynote researchers shared their scholarly stories with two hundred attendees to the Summit. I have never seen so many high school Ecuadorian kids and college undergraduates twitting science, reporting *in situ* the outcomes of each talk, interviewing the speakers, making science part of their youthful experiences, unafraid of asking the simplest, yet most important question: Why is evolution the foundation of all? Well, I replied, evolution offers the only naturalistic explanation about biological phenomena; it relies on empiricism and rationalism, on facts and scientific observations of rigor. Evolution is a reality that can be tested and replicated and, if us, humans, understand and embrace it, it offers the foundation of all scientific thinking. Evolution is not only about natural history, it is about appreciating our place in the universe and securing our descendants' future existence in healthy, evolving ecosystems.

The Summit was also about launching the Lynn Margulis Center for Evolutionary Biology, affiliated with USFQ, in celebration of a genuine seeker of nature's deep mysteries. Margulis (1938-2011) provoked fascinating controversies over the origin and evolution of cells, their nuclei and organelles, via symbiotic relationships among ancient life forms that apparently merged during the Earth's early past. Lynn co-proposed the hypothesis of GAIA (no relation to GAIAS) which suggested that the complex associations of all organisms in the planet engender a homeostatic balance, a harmonic coexistence responsible for life's perpetuity over eons. Lynn lived by this principle of fruitful association and influenced the academic careers of hundreds of naturalists; she was best friend to many, perhaps to most.

Chemistry Nobel Laureate (2009) Ada Yonath, Israel, delivered a brilliant talk on the evolution of ribosomes, organelles responsible for the assemblage of amino acids and, therefore, of proteins, the essence of Earthy life. Her charisma on stage paralleled the profound evolutionary relevance of the research. An exuberant journey inside the intimate confinements of molecules brought us, the audience, to imagine infinitude at the micro scale, where atoms harmonize with each other, where chemical properties and magnetic interactions resemble the delicate balance invoked by GAIA at the macro scale. I felt inner 'enlightment' when reassured by Ada that evolution is within us, in every particle of matter that makes who we are.

In my own talk, I addressed Evolution, Science, Pseudo Science and the Public's Perception of Reality; in essence, my concerns about the current patterns of low acceptance of evolution worldwide and the conflict between the belief in supernatural causation and the reality of scientific rationalism/empiricism. I concluded that coexistence between faith and science is illusory due to their inherent incompatibility and that the controversy over acceptance of evolution will continue, indefinitely, via alternating mild and intense antagonism. And while writing this piece, my research co-author and collaborator, Avelina Espinosa, alerted me about the upcoming debate The Origin and Evolution of Life: Is Galapagos a Detour? between Michael Denton, from the Center for Science and Culture (branch of the Discovery Institute —sponsor of the late doctrine of Intelligent Design) and Michael Shermer, editor of Skeptic Magazine. The duel is part of the Las Vegas gathering "freedomfest" a "festival —July, 2013— where free minds meet to celebrate great books, great ideas, and great thinkers..." I hope Shermer delivers the greatness of Galapagos to honor the reality of evolution and puts to rest Denton's efforts to smuggle the fictitious into science.

Suggested Readings and Resources

Chemistry Nobel Laureates 2009. http://www.nobelprize.org/nobel_prizes/chemistry/laureates/2009/.

Galapagos III World Evolution Summit.
http://www.usfq.edu.ec/eventos/evosummit/Paginas/default.aspx.

Galapagos Institute for the Arts and Sciences. http://www.usfq.edu.ec/programas_academicos/galapagos/Paginas/default.aspx.

Lovelock, J. E. & Margulis, L (1974). Atmospheric homeostasis by and for the biosphere: the Gaia hypothesis. *Tellus*, 26, 2–10.

Margulis, L. (1970). *Origin of eukaryotic cells*. New Haven, CT: Yale University Press.

Margulis, L. (1981). *Symbiosis in cell evolution*. New York, NY: WH Freeman.

Paz-y-Miño-C, G. & Espinosa, A. (2013). Attitudes toward evolution at New England colleges and universities, United Sates. *New England Science Public: Series Evolution*, 1, 1–32.

Paz-y-Miño-C, G. & Espinosa, A. (2013). The everlasting conflict evolution-and-science versus religiosity. In G Simpson & S Payne (Eds.), *Religion and ethics* (pp. 73–97). New York, NY: Nova Publishers.

Ribosome Structure and Function at Professor Ada E. Yonath website.

http://www.weizmann.ac.il/sb/faculty_pages/Yonath/.

Skeptic Magazine. http://www.skeptic.com/.

Keynote Speakers at the Galapagos III World Evolution Summit. http://www.usfq.edu.ec/eventos/evosummit/Paginas/Speakers.aspx.

Essay 5

CONSERVATION BEHAVIOR IN THE GALAPAGOS*

"…"I will… [give] an account of the extreme tameness of the [Galapagos] birds. This disposition is common to all the terrestrial species; namely, to the mocking-thrushes, the finches, wrens, tyrant-flycatchers, the dove, and carrion-buzzard. All of them often approached sufficiently near to be killed with a switch, and sometimes, as I myself tried, with a cap or hat. A gun is here almost superfluous; for with the muzzle I pushed a hawk off the branch of a tree. One day, whilst lying down, a mocking-thrush alighted on the edge of a pitcher, made of the shell of a tortoise, which I held in my hand, and began very quietly to sip the water; it allowed me to lift it from the ground whilst seated on the vessel: I often tried, and very nearly, succeeded, in catching these birds by their legs… It is surprising that they have not become wilder; for these islands during the last hundred and fifty years have been frequently visited by bucaniers and whalers; and the sailors, wandering through the woods in search of tortoises, always take cruel delight in knocking down the little birds…"

Charles Darwin, The Voyage of the Beagle, 1839

A great adaptation animals have is the ability to spot predators and avoid them; predators, in response, have to keep up with prey evolution. Elusive, poisonous, deceiving, confusing and difficult-to capture prey trigger countering responses in their predators, which have become, over millions of years, even more secretive, tolerant to noxious chemicals, smart and able to

---

* Essay first published in *The Conservation Behaviorist*, Vol. 5, 2007.

catch the fastest, most camouflaged and unpalatable prey. This "evolutionary arms race" is responsible for the enormous variety of prey and predatory traits existing in nature. But if predators are absent or rare, species mainly evolve behavioral repertoires to cope with the immediate struggle for life at the expense of behaviors that, if the environment changes by, for example, the arrival of unfamiliar enemies, could be essential for survival. "Maladaptive behavioral responses" or lack of appropriate action in presence of novel danger can drive animals into extinction, which nowadays is a concern in the Galapagos National Park, a protected volcanic archipelago located 600 miles of the coast of Ecuador.

Three to five million years of isolation have generated unique characteristics among the Galapagos endemic and resident species: cormorants have lost their ability to fly and now dive searching for fish in a turbulent ocean, marine iguanas feed on algae in the bottom of the sea shore, three-hundred pound giant tortoises digest cactus which spines cause no damage in their guts, lava lizards jump and tumble while capturing flies over the huge bellies of resting sea lions, bluefooted boobies and albatrosses saturate the ground with their nests and courtship displays, and various species of Darwin finches coexist by sharing seeds, insects and occupying distinct habitats near and above the ground. Although these tame fascinating creatures are fit for the Galapagos rocky landscape, human impact is threatening their survival. Introduced goats, cats, dogs, pigs, donkeys, and other animals, are destroying these fragile islands.

Settlers released domestic animals in the Galapagos five centuries ago. Feral goats now feed on the natural vegetation and stump on regrowths of the endemic flora, transforming the landscape from shrubby-dense plant cover to desert-like and eroded terrains. Cats and dogs prey on birds and reptiles, while pigs dig into nests and crash and eat the eggs of giant tortoises, marine and terrestrial iguanas. Because the Galapagos vertebrates are nonpoisonous or unpalatable and have had no phylogenetic exposure to mammalian enemies, predator avoidance behaviors develop very slowly.

After years of exotic animal invasions and other related human impacts on the islands (deforestation and farming), the Galapagos fauna seems to be losing this battle unless conservation plans are implemented. Captive breeding programs at the Charles Darwin Research Station (CDRS), in Santa Cruz Island, are giving positive results for the giant tortoises and terrestrial iguanas. Understanding their reproductive behavior and social organization together with research on feeding habits and dispersal patterns are helping biologists to reintroduce these species into the wild. But restoration requires entire habitat rehabilitation to secure viable populations. In Galapagos the solution is clear: feral nasty animals, such as goats, cats, dogs, pigs and donkeys must go! Who should go first? Goats, the nastiest; they are thriving in the Galapagos and overpowering tortoises and iguanas.

Creative hunting campaigns aimed at eradicating goats have been in practice over half a century. Most recently, goats have been trapped and released with radio tracking devices, which allow hunters to locate herds and gun them down from helicopters. Goats, however, do learn quickly to elude predators coming from the sky in flying vessels and soon turn erratically nocturnal and crepuscular, spread while being chased to congregate again and continue with their own survival. Confusion, dilution, odd prey effects, and other behavioral phenomena are taken into consideration by the CDRS to exterminate goats efficiently, yet the outcome is ambiguous. Hunting from aircrafts selects for fast, smart, even more opportunistic animals, plus is very expensive. The islands of Isabela, Santiago and Santa Cruz still have large populations of feral goats.

Galapagos -sanctuary of wildlife and United Nations World Heritage- faces its most challenging times ever. The archipelago is an example of both proper ecotourism/scientific management and acute human conflicts with nature. One hundred thousand visitors arrive in the National Park and Marine Reserve every year and follow strict rules and organized procedures to explore the islands. Scientists and students engage in research and environmental education programs (geology, biology, anthropology) sponsored by universities and nonprofit organizations. At the same time, the human population in Isabela, Santa Cruz and San Cristobal grows and enormous environmental pressure is imposed by farmers, fisheries and developers. Conservation problems continue and some worsen: fire ants, quinine trees, bees and rats invade the islands at an alarming rate, illegal sea cucumber harvesting and shark and tuna fishing propagate in protected waters, while powerful cruise corporations find their way into bringing entertainment ships to replace the local ecotourism enterprises. It is hard not to agree with Darwin and his disenchantment with the visitors of the nineteenth century who while "*...wandering through the woods in search of tortoises, always [took] cruel delight in knocking down the little [Galapagos] birds...*" Today's "wanderers" either driven by poverty or equipped with technology, credit and political influence are knocking down an entire World Heritage.

Brief History of the Galapagos

1535 Galapagos officially 'discovered' by the Bishop of Panama, Fray Tomás de Berlanga
1574 the Ortelius' map calls the archipelago 'insulae de los galapegos'
1500-1700s pirates, whalers and buccaneers control the islands
1832 Ecuador assumes jurisdiction of the 'Archipiélago del Ecuador'
1835 Charles Darwin visits the Galapagos Islands
1959 Ecuador creates de Galapagos National Park
1964 the Charles Darwin Research Station opens
1979 UNESCO declares Galapagos a World Heritage
1985 UNESCO declares Galapagos a Biosphere Reserve
2000s >100,000 visitors arrive in Galapagos every year

Suggested Readings and Resources

Darwin Online, The Zoology of the Voyage of HMS Beagle. http://darwin-online.org.uk/EditorialIntroductions/Freeman _ZoologyOfBeagle.html

Darwin Online, Journal of Researchers. http://darwin-online.org.uk/ EditorialIntroductions/Freeman_ JournalofResearches.html

Galapagos National Park. http://www.galapagospark.org/.

Galapagos, UNESCO's World Heritage. http://whc.unesco.org/en/list/1.

Idrovo, H. (2005). *Galapagos: Footsteps in paradise*. Quito, Ecuador: Libri Mundi.

Paz-y-Miño-C., G. (in press). Cognitive repertoire, cognitive legacies and collective reasoning: The next frontiers in conservation behavior. In Ken Yasukawa (Ed.) *Animal Behavior Three Volumes, Vol. III Emerging Research*. Santa Barbara, CA: ABC-CLIO Publishers.

Paz-y-Miño-C., G. (2007). Conservation behavior in the Galapagos. *The Conservation Behaviorist*, 5, 4-5.

Stewart, P. D. (2006). *Galapagos: The islands that changed the world*. New Haven & London: Yale University Press.

Essay 6

MAUNA KEA TELESCOPES TO SINK IN THE PACIFIC – HAWAII*

"…before the Dim End, all land telescopes will wear off and turn obsolete and… orbiting-the-planet observatories will likely replace them. Eventually, the shining domes resting on Mauna Kea will crumble while drifting away Northwest on their carrier, the late "Big Island." These magnificent pieces of engineering will sink in the Pacific… when the summit of Mauna Kea succumbs to erosion, hence following the drowning fate of the Hawaiian Islands."

Contemplating the night sky from the top of the largest volcano on Earth is spectacular. Indeed, on the Hawaiian Mauna Kea's slopes, thirteen astronomical observatories reach the universe through their light and radio-wave telescopes.

From base to summit, Mauna Kea (33,500 feet) raises taller than the Mahalangur-Himal Mount Chomolungma (29,000 feet) or, as renamed in the 1860s, "Everest." The Royal Geographical Society, responsible for the authoritarian christening, eagerly sought to immortalize fellow member George Everest, a Welsh prominent topographer, who in youth was Surveyor General of India (1830s-40s). Thus the centuries-old Tibetan tradition of honoring the marvelous Chomolungma with a beautiful native name (of course surreal "Mother Goddess") was lost.

But Mauna Kea, or "The White Mountain" (at times covered with snow), conserved its Hawaiian designation, traceable to the 11th century, and probably to the years 300-500, when Polynesians settled in the islands. The

---

* Essay first published in *The Standard Times* of New Bedford, September 18, 2012.

Mauna Wakea synonym, or "Mountain of the Deity Wakea," is probably a more recent adoption since mythical traditions regarding the landscape customarily develop after the descriptive word-stock.

Despite the appalling American and European business-men conspiracy to overthrow the Kingdom of Hawaii (1893), followed by a short-lived pseudo-sovereign republic (1894-98), an annexation as territory to the United States (1898), a granted statehood (1959), and a final Apology Resolution by the US Congress (1993) –which was co-signed by President Bill Clinton— for the "deprivation of the rights of Native Hawaiians to self-determination," the ancestral cultures reverberate in contemporary Hawaii.

Mauna Kea's volcanic foundations emerge from the mantle, deep below the Pacific tectonic plate. A hotspot fuels with magma the islands in the Hawaiian Ridge, and these "huge masses of interconnected rock" move slowly –at the speed our nails grow— toward the Northwest. The Big Island (0 to 600,000 years old), the newest formation and where Mauna Kea is located, rests on the hotspot, while the remaining smaller islands of the Maui cluster (1-2 million years old), Oahu (2-4 million years old) where Honolulu is situated, and the Kauai cluster (more than 5 million years old) continue to erode and sink away from the hotspot. The current archipelago is destined to disappear under the ocean and, if the hotspot remains active, to be replaced by new islands. Hawaii is an exemplar of the reality of an evolutionary tectonic process.

Regardless of the sporadic snowy peaks, the atmosphere over Mauna Kea is cloudless most of the year, free of high-elevation particle pollution and very dark with no artificial-nightlight influence, key factors for telescopic observations. And $2-billion in infrastructure and equipment have been brought to the summit by a dozen countries which work in partnership with the University of Hawaii, active manager of the "Astronomy Precinct" within the Mauna Kea Science Reserve.

The Mauna Kea telescopes look grandiose in close proximity. Their white and silver domes stand out at twilight while waking up for their night runs: in 1968-70, the University of Hawaii built the first two large observatories (UH 0.9-m educational telescope and UH 2.2-m telescope); followed by the NASA Infrared Telescope Facility (1979), Canada-France-Hawaii Telescope (1979), UK Infrared Telescope (1979), James Clerk Maxwell Telescope (1987), CALTECH Submillimeter Observatory (1988), Very Long Baseline Array (1992), Keck I and II (1993-6), Subaru (1997), the Gemini North (1999), and the Submillimeter Array (2002). These instruments explore outer space under optical, infrared, submillimiter and radio spectra.

I found much mysticism, however, intertwined with cosmological facts, at Imiloa, an impressive, beautifully colored educational facility, which is advertised as "part science center, part indigenous peoples museum" by the University of Hawaii at Hilo. Imiloa merges scientific knowledge and sacred traditions about "origins" as if they were empirically compatible, but they are not. We do know, for example, that neither the Hawaiian Islands nor humans were created, but that the former emerged from magma piercing its way out through the Earth's crust, while the latter evolved 180,000 years ago from African ancestors whose descendants, the Polynesians, arrived in Hawaii. And thanks to the Mauna Kean telescopes —plus two millennia of astronomy— we are certain that the universe evolved autonomously, independent from the invention of mythology, and that it will end when the last stars deplete their own fuel.

But before the Dim End, all land telescopes will wear off and turn obsolete and, if our species persists over a few more cosmic seconds in the time scale, orbiting-the-planet observatories will likely replace them. Eventually, the shining domes resting on Mauna Kea will crumble while drifting away Northwest on their carrier, the late "Big Island." These magnificent pieces of engineering will sink in the Pacific, as archeological relics, when the summit of Mauna Kea succumbs to erosion, hence following the drowning fate of the Hawaiian Islands.

SUGGESTED READINGS AND RESOURCES

Apology Resolution by the US Congress. http://www.gpo.gov/fdsys/pkg/BILLS-103sjres19enr/pdf/BILLS-103sjres19enr.pdf.

Caltech Submillimiter Observatory. http://cso.caltech.edu/.

Canada-France-Hawaii Telescope. http://www.cfht.hawaii.edu/.

Gemini Observatory. http://www.gemini.edu/.

James Clerk Maxwell Telescope. http://www.jach.hawaii.edu/JCMT/.

Keck I and II Observatories. http://www.keckobservatory.org/.

Kingdom of Hawaii. http://www.hawaiiankingdom.org/index.shtml.

Mauna Kea Science Reserve. http://www.hawaii.edu/maunakea/.

Macdonald, G. A., Abbott, A. T. & Peterson, F. L. (1983). *Volcanoes in the sea*. Honolulu, HI: University of Hawaii Press.

NASA Infra Red Telescope Facility. http://irtfweb.ifa.hawaii.edu/.

Royal Geographical Society. http://www.rgs.org/HomePage.htm.

Subaru Telescope. http://www.naoj.org/.

Submillimiter Array. http://www.cfa.harvard.edu/sma/.

University of Hawaii Hilo Educational Telescope. http://www.astro.uhh.hawaii.edu/NewEducationalTelescope.php.

University of Hawaii 2.2-meter Telescope. http://www.ifa.hawaii.edu/88inch/.

United Kingdom Infrared Telescope. http://www.jach.hawaii.edu/UKIRT/.

Very Long Baseline Array. http://www.vlba.nrao.edu/.

Ziegler, A. C. (2002). *Hawaiian natural history, ecology, and evolution.* Honolulu, HI: University of Hawaii Press.

Essay 7

BOSTON'S CHARLES HAYDEN PLANETARIUM*

"...Science is just a refined device for resolving ordinary curiosity and a powerful liberator of superstition. It is the subsistence kit to defeat re-emerging fundamentalism. And by recalling Giordano Bruno's and Galileo's past, as Santayana reasoned over the value of remembering history, we restate our right to learn the truth. This is why Boston's Charles Hayden Planetarium is more than a "light show," but a salutation to the triumph of scientific inquiry over blind belief..."

Giordano Bruno was not as fortunate as Galileo to opt for the imprisonment of his body in trade for thinking and writing about cosmology under house arrest. The Roman Inquisition orchestrated Bruno's "civil" burning in 1600, at age 52, for blasphemy and heresy, and for conjecturing that the sun was a star and that God and the universe were one and the same.

Galileo's fate, as philosopher, physicist and astronomer, was more honorable than Friar Bruno's. Although the Catholic church did find Galileo "vehemently suspect of heresy" in 1633, for defending Copernicus' proposal that the sun, not the Earth, was at the center of planetary orbiting, his precious mind could not be smoked by the clergy, nor smoldered by the populous, but rather retreated to concealed productivity. And so the church chaperoned Galileo's brainpower until age 77.

"*Those who cannot remember the past are condemned to repeat it*" proclaimed the Spanish-American logician George Santayana [or "*The One Who Does Not Remember History Is Bound To Live Through It Again*"].

---

* Essay first published in *The Standard Times* of New Bedford, August 17, 2011.

Indeed, recalling Bruno's or Galileo's tale is not pointless. It reminds us that exploring the cosmos can be obstructed by irrationalism, and that our worshipping for free science must continue to safeguard humanity's scientific legacies. It alerts us that the word "obscurants" was coined to denounce those who veil facts from the public.

And this brings me to the cosmic celebration role played by the renovated Charles Hayden Planetarium in Boston, but not before urging an equal revamp of its host, the Museum of Science itself.

Nitpicking aside, each "Undiscovered Worlds," "Explore the Universe," "Cosmic Collisions" or "The Sky Tonight" show is a tribute to centuries of space explorations, and a joyful remembrance that, despite Bruno's sadistic silencing and Galileo's confinement, the empirical truth prevailed over institutionalized ignorance.

The audience's amazement while watching the planetarium animations is contagious. A Zeiss Starmaster projector creates the illusion of space travel and it is fun to go along. Visual and sound effects impress human senses and the virtual take-off in a helicopter, from the museum's roof, while flying over Boston and metamorphosing into a spaceship that leaves Earth to immerse itself into galactic infinitude is magnificent.

Not surprisingly for an institution accountable for promoting science literacy, the planetarium has adopted clever marketing slogans: "*Discover the beauty and wonder of the night sky that has fascinated humanity for millennia*," or "*We have learned that our solar system is not alone in the universe, and we have had to redefine our understanding of planets and solar systems*," and my favorite "*Explore cosmic collisions, the hypersonic impacts that drive the dynamic and continuing evolution of the universe*."

And pro evolution aphorisms are vital for the 40% of New Englanders who still do not accept the reality of evolution and seek —together with 70% of their equally incredulous American counterparts— air-conditioned recreation at summer museums.

And in a free-market society where the Book of Genesis sells more than the Book of Reason, the Boston Museum of Science and its planetarium must compete not only with reputable national co-exemplars of proper public outreach, but with impostor "sister institutions," like the "creation museums" and "Genesis parks" emerging in the United States, where pseudoscience and intelligent design are smuggled in planetaria and aquaria format to depict foolishness, and so lure thousands of ticket buyers to harass plastic dinosaurs and inhale Eden.

But the Boston Museum of Science has just deployed an antidote to such pseudo-dinosauria frenzy: the exhibit "Dinosaurs: Ancient Fossils, New Discoveries," open until August 21 (2011), is another tribute to genuine science, and a co-effort with accredited institutions, including the American Museum of Natural History, Houston Museum of Natural Science, California Academy of Sciences, The Field Museum, and the North Carolina Museum of Natural Sciences.

Science is just a refined device for resolving ordinary curiosity and a powerful liberator of superstition. It is the subsistence kit to defeat re-emerging fundamentalism. And by recalling Giordano Bruno's and Galileo's past, as Santayana reasoned over the value of remembering history, we restate our right to learn the truth. This is why Boston's Charles Hayden Planetarium is more than a "light show," but a salutation to the triumph of scientific inquiry over blind belief.

Suggested Readings and Resources

Brewster, D. (1841). The martyrs of science, or, the lives of Galileo, Tycho Brahe, and Kepler. London, UK: John Murray.

Charles Hayden Planetarium. http://www.mos.org/planetarium.

Galilei, G. (2001 edition). Dialogue concerning the two chief world systems: Ptolemaic and Copernican. New York, NY: The Modern Libray.

Krauss, L. M. (2010). Cosmic evolution. Evolution, Education, and Outreach, 3, 193-197.

Rowland, I. D. (2008). Giordano Bruno: Philosopher/heretic. Chicago, IL: University Of Chicago Press.

Wootton, D. (2010). Galileo: Watcher of the skies. New Haven, CT: Yale University Press.

Yates, F. A. (1964). Giordano Bruno and the hermetic tradition. Chicago, IL: University Of Chicago Press.

Essay 8

ON THE WRONGLY CALLED "THE GOD PARTICLE"*

"…without mass, no atoms would exist, no galaxies or stars, no solar systems or planets with life, and no brains capable of thinking about it…."

Nobel laureate Leon Lederman affirms that the title of his 1993 book *The God Particle: If the Universe is the Answer, What is the Question?* offended two groups: those who believe in God and those who do not. But this is another astute —and a posteriori— marketing pronouncement. If true, Lederman and coauthor Dick Teresi, a science writer, would have disappointed 95% of all Americans (the 80% of believers and the 15% of seculars), the book's initial and major target audience.

As particle physicist, Lederman's intention with such an unfortunate and misleading heading —here I don't only blame the publishers for scrambling science with the supernatural to secure sales— was to precisely reach the populous obsessed with science fiction, more than with science facts, and discuss the potential existence of the Higgs boson (a subatomic particle), which experimental demonstration, as predicted for decades, could bring major understanding to the essence of matter.

This past 4th of July, 2012, the European Organization for Nuclear Research, CERN, celebrated with its own "fireworks," or highly energetic particle collisions, the discovery of a Higgs-like boson generated at the Large Hadron Collider (LHC), a magnificent underground instrument built to study the fundamental structuring blocks of all things.

---

* Essay first published in *The Standard Times* of New Bedford, July 12, 2012.

I visited CERN, last year [2011], located nearby Geneva, at the Swiss-Franco border. Its 17-mile circular accelerator speeds up, in opposite directions, subatomic "hadrons," either hydrogen nuclei or lead ions, which gain energy after consecutive laps. At the instant of collision, scientists recreate the conditions immediately after the Big Bang, resembling the first events in the existence of our 14 billion-year-old universe. CERN is shockingly impressive; its amazing technology and scale of engineering caused me profound joy.

Hadron collisions produce short-lasting minuscule particles difficult to detect, and the Higgs boson has been indeed elusive. Its existence was postulated in 1964, in separate articles published in Physical Review Letters by Robert Brout and Francois Englert, Peter Higgs (alone), Gerald Guralnik, Richard Hagen and Tom Kibble. But CERN seems to have found it or, as cautiously announced, "measured the products of its decay," thus inferring its existence.

To help us imagine this day of discovery, or presumption that the Higgs boson is real, in his 1990s book Lederman traces back the history of particle physics to 2,600 years ago; sparkled by the Greek philosopher Thales of Miletus, who wondered about the simplest forms of matter, continuing with Democritus of Abdera (c 400 BP), who not only coined the term atom ("uncuttable") but declared that "*…nothing exists except for atoms and empty space; everything else is opinion*," and ending with the 1993 cancellation, by the US Congress, of the Superconducting Super Collider project to be built in Waxahachie, Texas, and which would have surpassed the LHC at CERN with a 54-mile-diameter particle accelerator.

What are Higgs bosons? Remember that atoms consist of positively charged protons and electrically neutral neutrons, both occurring at a nucleus. Electrons cloud around the nucleus and are negatively charged. The entire atom package is kept together by electromagnetic forces. If a nucleus of hydrogen —the simplest known element which is essentially a proton— is accelerated and rammed against another proton an explosion occurs, which liberates subatomic particles. Physicists rely on a body of scientific knowledge, called the Standard Model, to theorize and explore experimentally —currently at CERN— the properties of such subatomic particles.

About 60 of these particles have been hypothesized and/or documented to exist, and scientists classify them as bosons, hadrons and fermions (for technical terminology visit CERN's Glossary, below). The Higgs is a boson and a crucial one to understand the properties of other elementary particles, for example, why some have mass and others, like the photons (components of light) don't. Without mass, no atoms would exist, no galaxies or stars, no solar systems or planets with life, and no brains capable of thinking about it. (Note, however, that Higgs-like particles are expected to account for only a fraction of the total mass of the universe). CERN asserts that the characterization of Higgs will provide "the final missing ingredient in the Standard Model" and guide us in the comprehension of the forces acting at the microscopic core of nature.

As for Lederman's book (I belong to the 15% of seculars who detest its heading and insertions of subliminal mysticism into the facts), the prose offers an enjoyable ride, rich in historicity, sarcastic humor —rare for a physicist— and fantasizing dialogs with Democritus, Lederman's imaginary physics peer. And to poise Lederman's enlightenment about particle physics and its ramifications to modern cosmology with the views of one of his contemporary elementary-particles colleagues, I recommend reading Victor Stenger's *God: The Failed Hypothesis* (2007), *Quantum Gods* (2009), and the latest *God And the Folly of Faith* (2012).

Suggested Readings and Resources

Carroll, S. (2012). *The particle at the end of the universe*. New York, NY: Penguin Group.

Englert, F. & Brout, R. (1964). Broken Symmetry and the Mass of Gauge Vector Mesons. *Physical Review Letters*, 13, 321–323.

European Organization for Nuclear Research CERN. The standard model. http://home.web.cern.ch/about/physics/standard-model.

Guralnik, G. S., Hagen, C. R. & Kibble, T. W. B. (1964). Global Conservation Laws and Massless Particles. *Physical Review Letters*, 13, 585–587.

Higgs, P. W. (1964). Broken Symmetries and the Masses of Gauge Bosons. *Physical Review Letters*, 13, 508–509.

Krauss, L. M. (2012). *A universe from nothing*. New York, NY: Free Press.

Physics Glossary by CERN. http://home.web.cern.ch/about/physics.

Stenger, V. J. (2007). *God: The failed hypothesis*. New York, NY: Prometheus Books.

Stenger, V. J. (2009). *Quantum gods: Creation, chaos, and the search for cosmic consciousness*. New York, NY: Prometheus Books.

Stenger, V. J. (2011). *The fallacy of fine-tuning: Why the universe is not designed for us*. New York, NY: Prometheus Books.

Stenger, V. J. (2012). *God and the folly of faith: The incompatibility of science and religion*. New York, NY: Prometheus Books.

The Large Hadron Collider at CERN. http://home.web.cern.ch/about/accelerators/large-hadron-collider.

Essay 9

A STATIONARY ARK ON THE ISLE OF JERSEY*

> "…Perhaps the ark metaphor can still be used, or resuscitated, as by Gerald Durrell, to inspire curiosity for investigating the frontiers of nature and to propose scientific solutions to the genetic erosion of increasingly small animal populations worldwide. Then there's the "full-scale" depiction of Noah's Ark at a 160-acre Genesis theme park to open in Kentucky in 2014, a $150 million investment that will intoxicate the minds of visitors with the idea that incest mating between two of a kind —just like Adam and Eve— can populate the Earth."

If the biblical story about the Ark were true, Noah's attempt to save animals from the Flood would have driven them into inbreeding depression.

Sentenced to incest, the descendants from the surviving sexual pairs — which themselves were allegedly confined to uncertainty in a drifting vessel— would have accumulated lethal combinations of genes and followed an inescapable vortex toward extinction.

Indeed, critically small populations of organisms store much less genetic wealth than the progenitors from whom they usually originate. Deprived from a healthy shuffling of paternal and maternal traits —a phenomenon known as recombination and that enriches the genome of creatures that mate — the offspring of repetitive kin procreations soon become unfit, fail to reproduce and gradually vanish.

The cosmos appears indifferent to this pain and suffering. Yet the laws of nature have steered the emergence of empathy and consciousness in the

---

* Essay first published in *The Providence Journal*, March 25, 2011.

universe via Darwinian selection. Self-aware apes, like humans, understand tragedy or joy when afflicted or delighted by them; empathic people can also infer when other animals experience distress or appeasement. But this assertion is incomplete, since the cognitive *Homo sapiens* is surrounded by a diverse mosaic of mental power in his close and distant phylogenetic relatives.

Intelligence has evolved gradually and in a continuum. And just like the fossil record, the pattern of human's "brained companions" is scattered after historical and always ongoing natural extinctions. Chimpanzees, bonobos, gorillas, orangutans, capuchin monkeys and feathered creatures such as ravens, jays and parrots possess and seek empathy. They are proof that the brain is another byproduct —not a goal— of favorable conditions for the assemblage of organic matter in a Goldilocks-zone planet, hospitable like ours, which has incubated life and later harvested intellect.

Empathy for each one of us and for other animals sometimes requires inspiration from the Ark's metaphor, but not from Noah's clumsy urgency to save provisions and secure self-subsistence in a landscape prone to fast inundation, but rather a well conceived and elegantly designed —under the scope of science— strategy for captive breeding and repatriation of endangered species.

I have a copy of *The Stationary Ark* (1976), with a handwritten note, "*For Guillermo, with best wishes, Gerald Durrell.*" But the first of Durrell's books, in which he referred to the concept of the Ark in the context of preservation and propagation of threatened animal populations, was *The Overloaded Ark* (1953), which was followed by *My Family and Other Animals* (1956). Both gave this Anglo-Irish animal fanatic, born in India in 1925, fame and the income to fund the Jersey Zoological Park (1958) and house his personal collection of exotic animals.

The Jersey Zoo became a world exemplar of how cutting-edge research, conducted at a tiny 30-acre patch of land in the English Channel Islands, off the French coast of Normandy, could lead to an international effort and successful captive breeding and reintroduction of the Mauritius kestrel and the Mongolian Przewalski's horse — evidence that zoos can reverse the course of some human assaults on nature.

Durrell adopted the flightless Dodo —a 3-foot-tall and 40-pound archaic pigeon endemic to Mauritius, in the Indian Ocean, which became extinct in the 17th Century— as symbol of the Jersey Wildlife Preservation Trust, JWPT (nowadays Durrell Wildlife Conservation Trust, DWCT), founded in 1963. The trust sponsors conservation of rare and endangered mammals, birds, amphibians and reptiles. The Wildlife Preservation Trust International (WPTI) was later established in the United States, in 1971, as counterpart benefactor of JWPT, and evolved into EcoHealth Alliance.

Today, the Stationary Ark not only oversees conservation research of 200 species of animals in peril, but inspires empathy for wildlife among "educated Noahs" at a mini-university established at Les Noyers, and adjacent to the zoo. It was there where, in 1987, Durrell autographed the book I have, while I was student at the International Training Centre (nowadays Durrell Conservation Academy, below) for captive breeding of endangered species; 2,700 graduates from 128 countries have been educated at the center since 1978, when it opened. —Durrell died at age 70.

Perhaps the ark metaphor can still be used, or resuscitated, as by Gerald Durrell, to inspire curiosity for investigating the frontiers of nature and to propose scientific solutions to the genetic erosion of increasingly small animal populations worldwide. Then there's the "full-scale" depiction of Noah's Ark at a 160-acre Genesis theme park to open in Kentucky in 2014, a $150 million investment that will intoxicate the minds of visitors with the idea that incest mating between two of a kind —just like Adam and Eve— can populate the Earth.

Suggested Readings and Resources

Durrell Conservation Academy. http://www.durrell.org/Training/.
Durrell Wildlife Conservation Trust. http://www.durrell.org/.
EcoHealth Alliance. http://www.ecohealthalliance.org/.

Essay 10

On Whales and a Whaling Museum*

"...Perhaps all cities should first be explored by visitors through the eyes of museums, and later be walked and valued for their details. The Whaling Museum of New Bedford is such a shepherd; it brings sight to the modern "Ishmaels" who can come to discover the shared ancestry between humans and whales or to understand the legendary madness connecting whalers to whales."

To be reassured that evolution is true one just needs to visit the New Bedford Whaling Museum.

Three mounted skeletons of a 50-foot North Atlantic right whale, a 40-foot humpback, and a 70-foot juvenile blue can impress anyone curious to compare human bones to those of whales. The vertebrae are identical in shape, the rib cage is a magnified version of a human's, and the forelimbs are shortened into appendices like "fins," each with digits. Only vestigial hips remain. The legs have disappeared during 35 million years of "sea galloping," thus passing on to the tail the job of thrusting the animal.

The skulls of whales differ much from those of primates, like ours, but every component is present. The orbits and cheek bones surround tiny eyes in respect to the large head, and the rostrum or face —made of the upper maxilla and nasal bones— is conspicuously protruded to meet the jaw, which, in the baleen whales like the trio above, has no teeth, only tough skin to hold up corneous plates made of hair-protein, evidence of ancestral furred relatives. Such whales gulp water to sieve shrimp, squid or fish.

The apparent lack of forehead is remarkable in cetaceans (whales and dolphins). During evolution, the frontal bone retreated toward the back of the

---

* Essay first published in *The Standard Times* of New Bedford, September 30, 2010.

skull and, in some species, created a boat-shaped cavity which stores wax, and the museum exhibits a fourth magnificent specimen to account for this, a toothed sperm whale. This 50-foot skeleton emerges over a silhouette of the whale's flesh sketched on the floor next to a giant squid, its favorite prey.

Herman Melville's 1851 edition of *Moby-Dick* rests ahead this specimen, on a glass plinth, almost challenging the beast to resume the chase, to dive into the old book and confront Ahab's obsession again — that of "wild vindictiveness" against Moby Dick for having reaped away the captain's leg— to bear harpoons piercing its back, to smite the Pequod and drown its crew, and to spare only Ishmael so that he drifts on a coffin —carpentered for his ill-with-sweatings friend Queequeg— and survives to narrate how the whale defeated all.

"Be ye also ready: for in such an hour as ye think not, the Son of man cometh," is encrypted on the cenotaph of "Captain WM. Swain, Master of the Christopher Mitchell of Nantucket … [who] after fastning to a whale, was carried overboard by the line, and drowned [on] May 19th, 1844." This dramatic epitaph on marble at the Seamen's Bethel —built in 1832 and still upright a few steps away from the museum— is analogous to those that daunted Melville while attending services in the early 1840s.

In his romantic novel, Melville imagined the pulpit of Seamen's Bethel as the bow of a whaler's ship, from which Father Mapple sermonized: "…Beloved shipmates… what is [the] lesson that the book of Jonah teaches? …the sin was in his… disobedience of the command of God… by seeking to flee from him… He [thought] that a ship… will carry him into countries where God does not reign… A dreadful storm [came] on, the ship [was] like to break… And Jonah… dropped into the sea… seething into the yawning jaws awaiting him; and the whale [shot]… his ivory teeth… Jonah prayed unto the Lord out of the fish's belly… the whale vomited out Jonah upon the dry land…"

The potential fury of whales and the obsession of whalers with killing them are historic. The Essex, which left Nantucket in 1819, was wrecked a year later by a sperm whale. The survivors ate the corps of shipmates while sailing small whaleboats 3,000 miles back to South America, ironically avoiding the Marquesas Islands, only 1,200 miles west from the Essex's sinking waters, where "cannibals could devour them." And the museum displays panels with estimates of whale massacres during the 18th through 20th centuries: 1 million sperm whales, 384,000 blues, 275,000 humpbacks, 92,000 bowheads, and 10,000 North Atlantic rights.

Whales were hunted for their oil, wax and baleen. Nothing was wasted. The collection of ornaments, utensils and capricious art on ivory, at the museum, accounts for that. Prosperity of the American colonies and the Industrial Revolution of the young republic (1820-1860) were fueled by the slaughtering of whales, and this ecocide must not be forgotten.

Perhaps all cities should first be explored by visitors through the eyes of museums, and later be walked and valued for their details. The Whaling Museum of New Bedford is such a shepherd; it brings sight to the modern "Ishmaels" who can come to discover the shared ancestry between humans and whales or to understand the legendary madness connecting whalers to whales.

Suggested Readings and Resources

Dolin, E. J. (2007). *Leviathan: The history of whaling in America*. New York, NY: W. W. Norton & Company.

New Bedford Whaling Museum. http://www.whalingmuseum.org/.

Philbrick, N. (2000). *In the heart of the sea: The tragedy of the whaleship Essex*. New York, NY: Penguin Books.

Smithsonian National Museum of Natural History. Ocean portal: media gallery video. http://ocean.si.edu/media-archive/video.

Videoclip, film Moby Dick 1956, Orson Welles as Father Mapple. http://www.youtube.com/watch?v=Mog0W6Jwj0Q.

Essay 11

DENYING ROME, THE EXQUISITE COLOSSEUM AND EVOLUTION*

"...denying the legacy of evolution becomes equivalent to rejecting one's kin or the deep DNA ancestry that connects us to worms and cyanobacteria; it is like capriciously ignoring the exquisite ruins of the Roman Colosseum which still stand, only 2,000 years young, on a 4.6 billion-year-old Earth."

History is evolution and evolution is history.

Charles Darwin felt enthralled when discovering the benefits of earthworms to archeology. In his 1881 book, *The Formation of Vegetable Mould Through the Action of Worms*, he dedicated a chapter to "the part which worms have played in the burial of ancient buildings." Darwinianly, he wrote "*...archeologists are probably not aware how much they owe to worms for the preservation of... ancient objects*."

Indeed, while tunneling in the dirt, earthworms soften the substrate and hide artifacts that otherwise would remain on the surface. Iron arrowheads, probably from the Battle of Shrewsbury in 1403, were, according to Darwin, preserved in this fashion by the aid of worms.

Superficial archeological ruins at Abinger (130-360s AD), Chedworth (350s AD), Brading (330s AD), and other vestiges of Roman occupation to Britannia have been concealed by the burring labor of earthworms. But Darwin clarified that "*...the enormous beds of rubbish, several yards in thickness, which underlie many cities, such as Rome, Paris, and London... have not been... acted on by worms*."

---

* Essay first published in *The Standard Times* of New Bedford, August 7, 2010.

As much as archeological information has been trapped underground — from earlier times located close to the surface to ancient epochs hidden deep— biological history is also preserved in the geological profile, from the Holocene (today) to the Cambrian (550 million years ago), when biodiversity fossilized vastly, and to the early Archaean (3.5 billion years ago), when colonial cyanobacteria carved rocks.

And as much as erosion disfigures the archeological record, the biological testimony is fragmentary to scientists. Both archaeological and biological treasures require ideal conditions for preservation. The majestic Roman Colosseum , for example, has battled gravity, rain and wind for two millennia.

Similarly, human bones from the Upper Paleolithic still remained identifiable to paleoanthropologists after 35,000 years of entombment in Abri de Cro Magnon, in Southern France. Like the Roman Colosseum, the aged fossils had lost essential features, but they existed and were (are) real.

It is here, at the point of recalling the dawn of Rome and the death of life forms in the process of leaving progeny —like the Cro-Magnons— where human logic faces the test. Denying biological evolution parallels with repudiating history. The antecedents of the Roman Empire are connected to its people, whose prosperous culture, technological pride and understanding of government were linked, historically, to simpler beginnings of the human condition, when hunter-gatherers strategized daily survival.

No rational citizen of the contemporary world would challenge the existence of ancient Rome, but 40% of Americans, 18% of the British, and 20% of Italians, among residents of 34 other countries where public acceptance of evolution has been polled, think evolution is false. This discrepancy between wide acceptance of Roman history and selective rejection of life's past resides in the extra scrutiny imposed on the latter by religion.

In the United States of America, negative attitudes toward evolution correlate with three variables: religious beliefs, pro-life beliefs and political ideology. Conservative Republicans accept evolution less than progressive liberals and independents (30 vs. 60%, respectively) and frequency of religious practices is associated negatively with acceptance of evolution (24% among weekly church-goers vs. 71% for seldom or never, see links to polls below).

Only science education improves evolution literacy; in fact, public acceptance of evolution in the United States increases from the high-school level (21%), to the some-college (41%), college-graduate (53%) and post-graduate levels (74%), reaching the highest among university professors (97%). Young Earth Creationists, who believe that humans were designed in the present form within the last 10,000 years, coincide with the views of the least educated population of adolescents in their teens.

In a May 2010 assessment of 35 universities and colleges in the progressive New England states, where public acceptance of evolution is the highest in the nation —only 59%— my laboratory documented that 97% of the faculty vs. 78% of the students accept evolution, and that 82% of the faculty vs. 58% of the students think that evolution is true. Notably, 91% of the professors admit to being "very concerned" or "somehow concerned" about the controversy of evolution vs. creationism and its implications for science education.

As much as accepting the reality of the Roman Empire, or Darwin's observations of the burring effects of worms on archeological remains, or the authenticity of the 17,000-year-old paintings by Cro-Magnons in the caves of Altamira, La Pasiega or Lascaux, admitting the veracity of biological history is a civilization's responsibility.

Otherwise, denying the legacy of evolution becomes equivalent to rejecting one's kin or the deep DNA ancestry that connects us to worms and cyanobacteria; it is like capriciously ignoring the exquisite ruins of the Roman Colosseum which still stand, only 2,000 years young, on a 4.6 billion-year-old Earth.

SUGGESTED READINGS AND RESOURCES

Paz-y-Miño-C, G. & Espinosa, A. (2013). The everlasting conflict evolution-and-science versus religiosity. Pp. 73-97. In Simpson G & Payne S (eds). *Religion And Ethics*, NOVA Publishers, New York.

Paz-y-Miño-C, G. & Espinosa, A. (2013). Attitudes toward Evolution at New England Colleges and Universities, United States. *New England Science Public: Series Evolution*, 1, 1-32.

The Gallup Poll. (2007). Majority of Republicans Doubt Theory of Evolution. http://www.gallup.com/poll/27847/Majority-Republicans-Doubt-Theory-Evolution.aspx.

The Gallup Poll. (2008). Evolution, creationism, intelligent design. http://www.gallup.com/poll/21814/Evolution-Creationism-Intelligent-Design.aspx.

The Gallup Poll. (2009). On Darwin's Birthday, Only 4 in 10 Believe in Evolution. http://www.gallup.com/poll/114544/xDarwin-Birthday-Believe-Evolution.aspx.

The Gallup Poll. (2010). Four in 10 Americans believe in strict creationism. http://www.gallup.com/poll/145286/Four-Americans-Believe-Strict-Creationism.aspx.

The Gallup Poll. (2012). State of the States: Mississippi is most religious US state. http://www.gallup.com/poll/153479/mississippi-religious-state.aspx#1.

The Pew Global Attitudes Project. (2007). World publics welcome global trade but not immigration. Washington D. C.
http://pewglobal.org/files/pdf/258.pdf.

The Pew Research Center For The People & The Press. (2005). Public divided on origins of life: religion a strength and weakness for both parties. Washington DC.
http://people-press.org/reports/pdf/254.pdf.

The Pew Research Center For The People & The Press. (2009). Scientific achievements less prominent than a decade ago: public praises science; scientists fault public, media. Washington DC. http://people-press.org/reports/pdf/528.pdf.

The Pew Forum On Religion & Public Life. (2008). US religious landscape survey – religious beliefs and practices: diverse and politically relevant. Washington DC.
http://religions.pewforum.org/pdf/report-religious-landscape-study-full.pdf.

Essay 12

Lisbon's Lesson: Honor The Value of Discovery*

"...Lisbon taught us a lesson: its beauty and history, museums and palaces, cathedrals and monuments all honored the value of discovery, the irrefutable foundation of true civilizations."

There is a connection between Portugal's cultural and historical commitment to explore the unknown and what just happened at the XIV European Society for Evolutionary Biology (ESEB) meeting held in Lisbon August 19 to 24, 2013. But a preamble is merited here before I address the conference's substantial outcomes.

Portugal's and Spain's leadership during The Age of Discovery (1500s-1600s) is undeniable. The Treaty of Tordesillas, signed in 1494, aimed at sharing between both kingdoms the geopolitical control of the world, as much as it could be explored, conquered and, inevitably by post-Crusade-invasion practices, Christianized. And so it was.

Exploration and discovery did nurture Europe's curiosity for sighting the "planet's final frontiers" for two centuries (XVI-XVII), starting with the uncertainty of Earth's shape and its implications for circumnavigation. But trade and profit were the vested motivators for the monarchies to "globalize" their understanding of the world which, 500 years ago, was not even conceived as a globe.

In a classical Type One error effort —to use modern science terminology— Christopher Columbus, a Genoese explorer sponsored by Spain, failed at arriving to Asia via the Pacific, and instead bumped into

* Essay first published in *The Standard Times* of New Bedford, September 23, 2013.

unfamiliar terra firma in 1492. Columbus was conceptually wrong and died, in 1506, unaware of the mistake, but his maritime adventure brought, nonetheless, unprecedented wealth to Europe.

The Portuguese Vasco da Gama tested with success, during 1497-1499, an alternative proposal: that India could be reached if sailing around Africa, relying, of course, on the Earth's roundness, plus novel technology, instrumentation, and vessel design.

In retrospect, Columbus and da Gama quests seeded today's world interconnectedness. But it was science inspiring mere pursuit of knowledge — equivalent to research programs— which led to the prosperity later harvested. Both Columbus and da Gama thought the former arrived in the "West Indies." Yet, it took additional expeditions (1499-1504), by cartographer Amerigo Vespucci, from Florence, to conjecture the existence of an entirely new continent in the Pacific, a major "paradigm shift" not unusual in science considering it relies on seeking the truth via skepticism.

And science per se, materialized for its intrinsic significance, curiosity-driven and respected for advancing knowledge and debunking myth —rather than for amassing fortune when its applications expedite income for entrepreneurs— was the spirit of 1500 international delegates gathered at ESEB 2013.

Lisbon was ideal for a conference about ancestry and change, legacy and improvement, the essence of evolutionary biology. At 34 symposia, 74 plenary talks, 360 speakers, and authors of 900 posters discussed genetic and non-genetic (cultural) inheritance of traits, animal behavior, mechanisms of species recognition to avoid hybridization, natural and sexual selection, host-parasite interactions, human evolution, aging and senescence, emergence of drug resistance, conservation of wildlife, online resources and quantitative simulations to teach evolution, and climate-change impacts on ecological and evolutionary processes.

A sense of "*fundamental research is what matters, not the sheer application of science for revenue*" resounded during the conference. The concern that funding for basic science is scarce worldwide, the disinterest among benefactors in sponsoring "why questions" in studies, and rather favoring the "*how much return will that generate for the industry, the patenting system, the biolabs, the administrative overheads*," and the uncertainty about the future of exploring ultimate queries —the reason for science's existence— were at the heart of small talking during the conference.

But ESEB 2013 was not alone in this respect. Earlier this summer, US researchers had specifically addressed the importance of sponsoring significant investigations. The Animal Behavior Society (ABS), for example, organized at its 50th anniversary meetings in Boulder, Colorado, July 28 to August 1st, the discussion "Time to Step Up! Defending Basic Science," under the premise that behavioral research has been "ridiculed" and caricatured by elected officials as "wasteful government spending." Ironically, behaviorists are the "role models" who continue to inspire worldwide interest in science, and ABS provided a list of them: Edward O. Wilson, Richard Dawkins, Judy A. Stamps, John Maynard Smith, and William D. Hamilton, among 15 others.

Likewise, the American Society for Microbiology featured in Denver, Colorado, May 18 to 21, the President's forum "Curiosity-Driven Basic Research: Laying the Foundation for Discoveries and Application of the Future," where "*the critical importance of basic investigations and the need to articulate why discovery is so essential*" was the consensus. And it cannot be otherwise at times when trivialization of reality, fed by entertainment, belief in the supernatural, disrespect for education, and self confidence nourished by how much is in the pocket, rather than in the schooled mind, can lead the populous to applaud emptiness.

But Lisbon taught us a lesson: its beauty and history, museums and palaces, cathedrals and monuments all honored the value of discovery, the irrefutable foundation of true civilizations.

Suggested Readings and Resources

Cliff, N. (2011). Holy War: How Vasco da Gama's Epic Voyages Turned the Tide in a Centuries-Old Clash of Civilizations. New York, NY: HarperCollins Publishers.

Holmes, R. (2010). The Age of Wonder: The Romantic Generation and the Discovery of the Beauty and Terror of Science. New York, NY: Vintage Books.

Mann, C. C. (2012). 1493: Uncovering the New World Columbus Created. New York, NY: Vintage Books.

Mann, C. C. (2006). 1491: New Revelations of the Americas Before Columbus. New York, NY: Vintage Books.

Essay 13

Can We Forecast the Fall of Today's Empires?*

> "…Empires can indeed supplant one another. And the example of the quipu versus the European bookkeeping systems illustrates how the clashing of technologies can freeze in time a "clever-practical system of information storing," the quipu, and replace it with a more efficient written mathematical notation. But the vicious stamping of the Iberian bull over the Andean culture was atrocious."

History reminds us that empires can supplant one another…

The Inca Empire (1430s-1530s) relied on an ingenious system of numeric recording, the *quipus*, or serial knots tied in colored threads made of hair from South American camelids, and used by accountants — the Quipucamayucs — to convey information on the calendar, trade, tributes and census, to other quipu-literate authorities who decoded the "talking knots" or used them in arithmetic operations (see link to Khipu Database Project at Harvard University).

Messages encoded in ropes were carried by Chasquis, or runners, who delivered quipus from Cuzco, the capital of the Tawantinsuyu —the Inca Empire— to its four Andean provinces. The network of Inca roads impressed the *conquistadores*, including Francisco Pizarro, who mobilized his 168 men and a few horses through the highland trails to reach Cajamarca —nowadays Peru— and, in 1532, captured and later killed the last of the Sapa Incas, Atahualpa, whose 80,000 troops succumbed to the military superiority of the Spaniards.

---

* Essay first published in *The Standard Times* of New Bedford, December 13, 2010.

Ironically, smallpox was the major executor of the Amerindians, next to the brutality of the conquest and Christianization crusade to suppress and replace with Catholicism not only the Inca god, the Sun —at least a real star venerated for nourishing maize and tuber fields— but any trace of cultural identity, including the *quipu*, the *chasquis* and the sound of their *pututos* — large conch shells used as horn-trumpets to announce the arrival of a message— and the Quechua language.

In "Of Rage and Redemption: The Art of Oswaldo Guayasamin (1919-1999)," the Ecuadorian painter depicts an Andean condor subduing a Spanish bull, shockingly illustrating how Inca descendants still wear the scars of the European invasion to their land. The Inca civilization collapsed in the 1530s, and 500 years of misery shadowed the survival of Indian villages in the Andes of Ecuador, Peru, Bolivia, Chile and Argentina.

Besides diseases, oppression, imposed ignorance to secure dominance, and neglect after war, what factors drive the decline of sociopolitical assemblages? What processes make civilizations emerge?

In two fascinating books, *Guns, Germs and Steel* (1997) and *Collapse* (2005), Jared Diamond, professor at the University of California Los Angeles, theorized how warfare, immunity or susceptibility to pathogens, and technology have driven the evolution of archaic societies, and how depletion of resources determined the breakdown of prosperous communities.

But the sequential steps of societal evolution had not been explored quantitatively.

Using evolutionary biology, researchers from the University of Tokyo, University College London and University of Auckland, New Zealand, have plotted models of societal organization onto linguistic diversification trees of 84 Austronesian (Southeast Asian and Pacific) groups.

Thomas Currie and his collaborators have demonstrated that political complexity during island colonization, from Taiwan to Western Polynesia and Southeast Asia, raised gradually —and predictably. It did from leaderless societies to simple chiefdoms (one leader), complex chiefdoms (one leader over another) and states (hierarchical leadership), rarely skipping in-between levels of organization.

Collapses, however, did occur in jumps, from state to simple chiefdoms or leaderless levels, or from complex chiefdoms to unorganized aggregations.

Currie and co-authors think that their statistical analyses "move us beyond purely verbal arguments" about "history's broadest pattern" of societal growth and breakdowns, and offer a model for deeper digging into the predictability of cultural change.

If the encounter between the Tawantinsuyu and the conquistadores led inevitably to the disintegration of the Incas —due to the technological disparity between the New and Old Worlds— and if the European economies of the 15th to 17th centuries, like Spain, Portugal and France, later lost predominance when facing the emergence of England (18th to 19th centuries), can we forecast the fall of today's empires?

Would market rivalry alone push aside prevalent systems and replace them inescapably with competitors? Would human capital immersed in technological skillfulness suffice to substitute the wealthy illiterate societies?

Empires can indeed supplant one another. And the example of the quipu versus the European bookkeeping systems illustrates how the clashing of technologies can freeze in time a "clever-practical system of information storing," the quipu, and replace it with a more efficient written mathematical notation.

But the vicious stamping of the Iberian bull over the Andean culture was atrocious.

SUGGESTED READINGS AND RESOURCES

Currie, T. E., Greenhill, S. J., Gray, R. D., Hasegawa, T. & Mace, R. (2010). Rise and fall of political complexity in island South-East Asia and the Pacific. *Nature*, 467, 801-804.

Diamond, J. M. (1997). Guns, Germs, and Steel: The Fates of Human Societies. W.W. Norton & Co.

Diamond, J. M. (2005). *Collapse: How societies choose to fail or succeed.* New York, NY: Penguin Books.

Khipu Data Base Project at Harvard University. http://khipukamayuq.fas.harvard.edu/.

Essay 14

ALL HISTORY IS BLACK HISTORY*

"...It has often and confidently been asserted, that man's origin can never be known: but ignorance more frequently begets confidence than does knowledge: it is those who know little, and not those who know much, who so positively assert that this or that problem will never be solved..."

Charles Darwin, The Descent of Man, 1871

An African American blogger just settled an online dispute over human origins by asserting, with magnificent wisdom, that "*all history is Black History*."

His co-bloggers had rejected the evolutionary significance of a recent fossil discovery in Hadar, Ethiopia, of a 3.2 million year old foot bone belonging to *Australopithecus afarensis* —the famous Lucy's lineage— which has unveiled hints about the locomotion habits of early hominids. The clever blogger brought reason into a trivial exchange of brain gas among sponsors of one wrong over another belief about human ancestry. "*Deal with it*" —that is with our common African heritage— were his last gentle, yet shut-up devastating words.

The left fourth metatarsal bone described in Science magazine by professors Carol Ward, William Kimbel and Donald Johanson, the latter discoverer of the very *Lucy*, also in Hadar, in 1974, tells a compelling story: Lucy's relatives walked upright, a conclusion awaiting unequivocal evidence for almost four decades, and the tiny bone just provided it in its complex anatomy. The now famous AL-333-160 three-inch bone (in paleontology

---

* Essay first published in *The Standard Times* of New Bedford, March 2, 2011.

identification numbers are precious) resembles the bipedal human "arched" metatarsals —shock absorbers for "walker apes"— and differs from the quadrupedal chimpanzee's, bonobo's and gorilla's, which are more elongated.

Lucy, herself 3.2 million years old, still is the most famous, although not most significant, hominid fossil ever described. *Ardi*, a 4.4-million year old almost complete skeleton discovered in Aramis, Ethiopia, by Tim White in 2009, and *Toumai*, a 7-million year old hominid skull found in Chad by Michel Brunet in 2001, are Lucy's fair competitors.

Lucy was brought to celebrity status by her always calculating manager, Johanson, who named her after The Beatles song *Lucy In The Sky With Diamonds*, thus launching her popularity into textbooks, anatomical reconstructions, cartoons of a stocky yet friendly smiling "ape cousin," and full-body hairy depictions of her 3-feet tall "reconstructed flesh" at museums and exhibits of human origins.

For bloggers and alike who question the African origin of humans, of which Lucy's tale is only part of the story since she was also born to common ancestors between her family and ours, the enigma has long been solved. But the journey to acknowledge that "*we are all Africans*" has been tortuous.

In The Descent of Man, 1871, Charles Darwin reasoned: "*...In each great region of the world the living [animals] are closely related to the extinct species of the same region. It is therefore probable that Africa was formerly inhabited by extinct apes closely allied to the gorilla and chimpanzee; and as these two species are now man's nearest allies, it is somewhat more probable that our early progenitors lived on the African continent than elsewhere.*"

Darwin's inference was brilliant, but it took more than a century to be definitely embraced by scholars. This "Out of Africa" hypothesis has been corroborated by DNA evidence, physical anthropology, and specimens.

Indeed ancient forms of *Homo sapiens* evolved to anatomically modern humans entirely in the Ethiopian realm, about 200,000 years ago. And they walked out of Africa to populate the planet, which can be demonstrated with fossils and genetic pedigrees of indigenous peoples worldwide. Even human languages follow a pattern of geographical distribution consistent with a common origin in Northeastern Africa and nearby Mesopotamia.

More recently, the "Multi Regional" hypothesis of human origins, which counters the out-of-Africa postulate by conceiving an overall hybridization and integration of all world *Homo* species, including archaic forms, such as the Asian Peking Man, *Homo erectus*, and modern variants, such as the Northeast African and Southern European *Homo sapiens*, into a unified single *Homo sapiens sapiens*, has regained closer attention —after losing power during the

1990s— due to the discovery of shared genetic features between us and Neanderthals, a fairly modern human variety which became extinct 40,000 years ago.

The single origin or Out of Africa idea is nowadays well accepted among scientists. And even if the multi-regional hypothesis regains strength, all ancestral forms of hominids seem to coalesce to Africa. Disagreements among scientists do not invalidate science as believed by misinformed illiterate bloggers, who insist that there are not enough fossils to account for evolution or that the questions about human origins are far from being answered.

As much as our African American blogger so intuitively stated that all history is, ultimately, Black History, Darwin's own writings from 1871 continue to alert us about the essence of belief-based debates: "*...It has often and confidently been asserted, that man's origin can never be known: but ignorance more frequently begets confidence than does knowledge: it is those who know little, and not those who know much, who so positively assert that this or that problem will never be solved.*"

Suggested Readings and Resources

Eltis, D. & Richardson, D. (2010). *Atlas of the transatlantic slave trade*. New Haven & London: Yale University Press.

Johanson, J. & Edgar, B. (2006). *From Lucy to language*. New York, NY: Simon & Schuster.

Ward, C. V., Kimbel, W. H. & Johanson, D.C. (2011). Complete fourth metatarsal and arches in the foot of Australopithecus afarensis. *Science*, 331, 750-753.

Wells, S. (2007). *Deep Ancestry: Inside the Genographic Project*. Washington DC: National Geographic.

Wells, S. (2004). *The Journey of Man: A Genetic Odyssey*. New York, NY: Random House Trade Paperbacks.

Essay 15

AMERICAN EXCEPTIONALISM BUILT ON BACKS OF THE 99%*

> "...If a bailout is an "act of loaning or giving capital to a company, a country, or an individual that is in danger of collapsing, in an attempt to prevent ruin," then the exceptionalism that we treasure must be rescued —"bailed out"— by the prosperous untouchable class that amassed fortune over the exceptional labor of their compatriots..."

A sense of inner exceptional nature probably drives the existence of most individuals and seems to be a Darwinian adaptive trait. Tribal pride is not only ancestral but ubiquitous.

Overconfidence, or "*believing you are better than in reality, is advantageous because it increases ambition, morale, resolve, persistence or the credibility of bluffing, generating a self-fulfilling prophecy in which exaggerated confidence boosts up success*," write Dominic D.P. Johnson and James H. Fowler in their latest article, *The Evolution of Overconfidence*, published in Nature in September, 2011.

Johnson and Fowler remark that "*...populations become overconfident, as long as benefits from contested resources are large compared with the cost of competition. The fact that overconfident populations are evolutionarily stable in diverse environments may explain why overconfidence remains prevalent today, even if it contributes to hubris, market bubbles, financial collapses, policy failures, disasters and wars.*"

In "Democracy in America" (Volumes I, 1835, and II, 1840; link to Democracy in America Online, below), the French sociologist and historian

---

* Essay first published in *The Standard Times* of New Bedford, November 2, 2011.

Alexis de Tocqueville acknowledged and critically contemplated the originality and uniqueness of the 19th century emerging superpower, the United States. He called it "exceptional" for its generalized equality and constitutional democracy, and for its commercial habits and pragmatism.

In retrospect, his observations about American ideology, cultural cohesiveness and shared values were astute and visionary. But de Tocqueville also detected the underlying fabric of puritan thinking, the essence of our modern conservatism:

"*Almost all the sects of the United States are comprised within the great unity of Christianity, and Christian morality is everywhere the same. In the United States the sovereign authority is religious, and consequently hypocrisy must be common; but there is no country in the whole world in which the Christian religion retains a greater influence over the souls of men than in America, and there can be no greater proof of its utility, and of its conformity to human nature, than that its influence is most powerfully felt over the most enlightened and free nation of the earth [Alexis de Tocqueville].*"

The clause "American exceptionalism" was apparently used, depreciatively, by Joseph Stalin (1878-1953) while forecasting in the 1920s that the American wealth, technological pride and social fairness were temporary and fated to collapse, unless communism took over. Ironically, the Soviet system crumpled in the 1980s and its "restructuring" —Perestroika— led to political and economic reform by mimicking the market-based models of the West, which, paradoxically, became today's fiasco in a landscape of deregulated money-making opportunities that did not benefit —mainly— the people.

The idea of exceptionalism consolidated in the United States and abroad after the 1950s, based on unparalleled industrial transformations, scientific achievements, competitive labor and education. Reality surpassed the cliché of uniqueness and actually disproved one of de Tocqueville's assertions: that Americans' exclusive trading interests "*diverted their minds from the pursuit of science, literature and the arts*" and that, therefore, exceptionalism was evident mainly in socio-economic affairs.

But living up to exclusiveness has proved difficult, and fast excessive wealth has led to ignore the history that created the "extraordinary." It was judicious investment in rigorous education, basic science and technology that gave rise to the best universities in the world, the finest hospitals, unrivaled space explorations, magnificent natural history museums and ecologically managed national parks. Farms, factories, highway connectivity and urbanization prospered due to an agile economy that generated, in hindsight, short lasting bonanza.

And it was an outstanding work force, driven by the highest standards of performance, that enriched today's heartless, and sometimes brainless, "top 1%," the financial conservative elite who opposes science, mocks intellectuals and ridicules college education. The overconfident mega-wealthy who question the reality of human-induced climate change, reject evolution, blame vaccines for causing mental retardation or autism, diminish the importance of biodiversity, and oppose environmental protection and clean energy.

Yet, they claim the value of exceptionalism as their own and intend to renegotiate it for the upcoming elections, re-sell it to the voters —as in the profitable stock market— of course, without committing their assets, or taxes, to the continuation of the "extraordinary," thus neglecting that "uniqueness" took a century to be erected over the human capital of workers, artists, scientists, musicians, teachers, novelists or poets, who invested themselves, and fully, to harvest collective good.

If a bailout is an "act of loaning or giving capital to a company, a country, or an individual that is in danger of collapsing, in an attempt to prevent ruin," then the exceptionalism that we treasure must be rescued —"bailed out"— by the prosperous untouchable class that amassed fortune over the exceptional labor of their compatriots.

Suggested Readings and Resources

de Tocqueville, A. (1835/1840). Democracy in America. http://xroads.virginia.edu/~HYPER/DETOC/toc_indx.html.

Johnson, D. D. P. & Fowler, J. H. (2011). The evolution of overconfidence. *Nature*, 477, 417-420.

Sachs, J. D. (2011). *The Price of Civilization*. New York, NY: Random House Trade Paperbacks.

The Equality of Opportunity Project. http://www.equality-of-opportunity.org/.

Essay 16

Rejection of Science Threatens to Be Epidemic*

> "…There is a civic duty all citizens can exert to rectify the politics obscuring education: Cast our votes for candidates who sponsor proper schooling and support significant, not only profitable research. Escort out of office those who see fiction and facts compatible, or worship ignorance-based opinions as rightful views of equitable value to the empirical truth..."

Houston, we have a problem. And it has nothing to do with the explosion of an oxygen tank at 200,000 miles away from Earth that is threatening the lives of three astronauts cramped in a 1970s model of a spaceship, nor with an imminent meteor shower or solar radiation blast.

It is the rejection of science by elected officials and their constituents who, although privileged to grow up in a nation leading the most important quest of all, that of superb education and cutting-edge discoveries at prestigious universities, now dismiss the value of knowledge and of scientific realities essential to our existence.

And it is not only evolution that is rejected (my favorite topic, as a biologist and university faculty committed to education) but specifically space explorations, climate change research, stem cell studies, cloning and vaccinations.

The opposition resides, at times, on costs, a legitimate reason when prioritizing funding for billion-dollar projects like NASA's Apollo (1960s-1970s), Shuttle (1980s-2000s) or International Space Station programs (1990s-

---

* Essay first published in *The Standard Times* of New Bedford, January 11, 2012.

2000s), but the resistance to the other fields is dubious under the economic justification (climate change) or relies on puritan thinking rather than on pro-health sincerity (stem cells, cloning and some vaccines).

It is impossible to honor knowledge when a nation's admiration for it vanishes, when its students rank 25th in math and 17th in science worldwide, or when the youth of 14 other countries reads more than our own, or when the term "professorial" becomes an insult or a reprehensible trait in a public figure, as if the sophistication gained by formal education were a malady that must be eradicated to ensure "equality" and fair share of omnipresent unawareness of all issues.

In the quest for attaining this absurd egalitarian recognition to low and high standards, the American colleges and universities have fallen into a pervasive grade-inflation. According to the Teachers College Record, a solid C was the most popular grade in the 1940s (35%), followed by B (below 35%), then by A (15%), and finally by D and F (above 10 and 5%, respectively). In the 1970s, the Vietnam War draft triggered a proliferation of A's, which surpassed 30%. Today the A grade is fashionable (over 40%) and the "uncool" C is granted to only 15% of all college students —a situation accentuated at private institutions.

But trivializing education can be suicidal in a competitive job market where only earning a bachelor's degree would keep a U.S. worker out of poverty. Indeed, education pays by reducing unemployment and rising income, and The Bureau of Labor Statistics has examined these trends: 15% of those without a high school diploma remained unemployed during 2010 and, if employed, they earned only $450 per week. Those who graduated from college reduced their unemployment to 5% and, if employed, earned more than $1,000 weekly. Only holders of master's degrees and above —professional and doctorate degrees— secured a job more than 95% of the time and earned beyond $1,300 per week.

Although the crisis in the current educational system is multi factorial and complex, there is a civic duty all citizens can exert to rectify the politics obscuring education: Cast our votes for candidates who sponsor proper schooling and support significant, not only profitable research. Escort out of office those who see fiction and facts compatible, or worship ignorance-based opinions as rightful views of equitable value to the empirical truth.

What can we achieve if public officials and their electors treasure education? An overwhelming support to science and reason; in fact, if citizens advance from holding a high school diploma to graduating from college, societal concurrence with major research topics will increase: space explorations from 50 to 70%; climate change from 45 to 58%; acceptance of evolution from 21 to 74%; embryonic stem-cell research from 51 to 71%; therapeutic cloning from 64 to 73%; and childhood immunization from 79 to 91% (data from Space Policy Journal, Gallup, Pew Research Center, Georgia Department of Human Resources and Coalition for the Advancement of Medical Research).

And going back to Houston, and specifically to NASA, our "competent sentinel" when cosmic menace approaches Earth, its terrestrial problem is now to secure $19 billion during 2012 —about 0.6% of the $3 trillion federal budget— and continue with the programs: science (planetary, astrophysics), aeronautics, space technology, exploration, space operations and education, the latter alone worth $138 million. Indeed, world-quality research and education can be expensive; is someone willing to try ignorance?

Suggested Readings and Resources

Bureau of Labor Statistics. Education pays: Unemployment and median weekly earnings as function of education.

http://www.bls.gov/emp/ep_chart_001.htm.

NASA Science Program benefits: NASA's positive impact on society. (2012). http://www.nasa.gov/50th/50th_magazine/benefits.html.

Rojstaczer, S. & Healy, C. (2012). Where A is ordinary: The evolution of American college and university grading 1940-2009. *The Teachers College Record*, 114, 1-23.

Essay 17

NEW ENGLAND PROFESSORS ACCEPT EVOLUTION, BUT THEY ARE RELIGIOUS*

"...People do not "believe" in evolution; we either accept it, or doubt about it, or reject it. But the reality of the evolutionary process continues regardless of our cognitive awareness or position about it. Evolution is true..."

Together with my collaborator, Dr. Avelina Espinosa, professor at Roger Williams University in Rhode Island, we have uncovered the patterns of acceptance of evolution among university professors in New England, and the results are both fascinating and startling.

A cultural assumption has been that scholars are supportive of science and remain distant from belief-based perspectives regarding the natural world. Is this factual?

We surveyed 244 faculty —90% Ph.D. holders in 40 disciplines at 35 colleges and universities widely distributed geographically in Connecticut, Maine, Massachusetts, New Hampshire, Rhode Island and Vermont. Our study was recently published online in Evolution, Education and Outreach, and the hard-copy report will appear in the March 2011 issue of the journal.

Why New England? The first shocking fact that triggered our interest in studying the Northeast of the United States was that, back in 2005, the Pew Research Center for the People and the Press had documented that only 59% of New Englanders accept evolution, the highest score nationwide, and that the overall regional acceptance of evolution in the United States was even more distressing: 57% in the Northwest, 45% in the Midwest, and 38% in the South.

---

* Essay first published in *The Standard Times* of New Bedford, January 15, 2011.

More alarmingly, in 2006, the United States ranked 33rd among 34 other countries where acceptance of evolution was assessed, in contrast to Iceland, Denmark, Sweden, France, Japan and the United Kingdom, top in the list, where 75 to 85% of adults accept evolution.

Our study revealed that 91% of the New England professors were very or somehow concerned about the controversy of evolution versus creationism versus "intelligent design" and its implications for science education. In fact, 96% of them supported the exclusive teaching of evolution in science classes and a 4% minority favored equal time to evolution and creationism (the latter declared unconstitutional —in public schools— by the Supreme Court: Edwards v. Aguillard 1987). And 92% of the faculty perceived intelligent design as not scientific and as proposed to counter evolution, or as doctrine consistent with creationism.

Although 92% of the professors thought that evolution relies on common ancestry —or that organisms can be traced back in time to ancestors that reproduced successfully and left descendants— one in every four faculty did not know that humans are apes, or relatives of primates. Worse, 30% of the faculty were Lamarckian, or believed in the inheritance of acquired traits during an organism's lifetime, like longer necks, larger brains, or resistance to parasites, which are passed on to the progeny, a hypothesis rejected a century ago.

We asked the professors if faith in God is necessary for morality, if religion is important in their lives, and if they pray. Only 5% agreed with the need of a God to secure proper social behavior, but 30% considered religion to be very important in their daily existence, and 17% confessed to pray daily.

The one-third of the faculty who thought that religion is important in their lives was comparable to the 33% of American scientists who admit to believe in God (Pew Research Center, 2009), but differed from the 12% of "professional evolutionary scientists" —members of the North American, European, United Kingdom, and other countries' National Academies of Sciences (American Scientist, 2007)— and particularly the 7% of members of the United States National Academy of Sciences who believe in a personal God (Nature, 1998).

Indeed, most international scientists and the elite of the United States researchers are not religious.

Why does acceptance of evolution matter? Because public acceptance of evolution in the United States (about 40%) correlates with support to: (1) proper science education in public schools; (2) science and technology as essential components of development and prosperity; and (3) rationalism and freedom of thought, all indisputable ingredients for a thriving society.

And it also matters because only the highly educated university professors of New England — hopefully of the nation — have levels of acceptance of evolution (97% according to our study) comparable to or higher than the ordinary public in other industrialized countries of Northern and Western Europe.

Because attitudes toward evolution correlated positively with understanding of science and negatively with religiosity and political ideology, aspects examined in our study, we concluded that science education combined with vigorous public debate should suffice to increase acceptance of naturalistic rationalism and decrease the negative impact of creationism and intelligent design on collective evolution literacy.

Suggested Readings and Resources

Graffin, G. W. & Provine, W. B. (2007). Evolution, religion and free will. *American Scientist*, 95, 294-297.

Paz-y-Miño-C, G. & Espinosa, A. (2013). The everlasting conflict evolution-and-science versus religiosity. Pp. 73-97. In Simpson G & Payne S (eds). *Religion And Ethics*, NOVA Publishers, New York.

Paz-y-Miño-C, G. & Espinosa, A. (2013). Attitudes toward evolution at New England colleges and universities, United States. *New England Science Public: Series Evolution*, 1, 1-32.

Paz-y-Miño-C, G. & Espinosa, A. (2012). Introduction: Why people do not accept evolution? *Journal of Eukaryotic Microbiology*, 59, 101-104.

Paz-y-Miño-C, G. & Espinosa, A. (2012). Educators of prospective teachers hesitate to embrace evolution due to deficient understanding of science/evolution and high religiosity. *Evolution: Education & Outreach* 5: 139–162.

Paz-y-Miño-C., G. & Espinosa, A. (2011). New England faculty and college students differ in their views about evolution, creationism, intelligent design, and religiosity. *Evolution, Education, and Outreach*, 2, 323-342.

Paz-y-Miño-C., G. & Espinosa, A. (2009). Assessment of biology majors versus non-majors views on evolution, creationism and intelligent design. *Evolution, Education and Outreach*, 2, 75-83.

Paz-y-Miño-C, G. & Espinosa, A. (2009). Acceptance of evolution increases with student academic level: a comparison between a secular and a religious college. *Evolution, Education, and Outreach*, 2, 655-675.

US Supreme Court: Edwards v. Aguillard (No. 85-1513). (1987). http://www.law.cornell.edu/supct/html/historics/USSC_CR_0482_0578_ZS.html

The Pew Research Center For The People & The Press. (2005). Public divided on origins of life: religion a strength and weakness for both parties. Washington DC.

http://people-press.org/reports/pdf/254.pdf.

The Pew Forum On Religion & Public Life. (2008). US religious landscape survey – religious beliefs and practices: diverse and politically relevant. Washington DC.

http://religions.pewforum.org/pdf/report-religious-landscape-study-full.pdf.

The Pew Forum on Religion and Public Life. (2010). US religious knowledge survey. http://www.pewforum.org/U-S-Religious-Knowledge-Survey-Who-Knows-What-About-Religion.aspx.

Essay 18

MASSACHUSETTS GETS AN A- IN SCIENCE STANDARDS*

> "…If schools and teachers stick to a "C average" state of science standards, what quality of education are we really offering? If state science standards guide science education, how do we escape from this loop of poor standards and the expectation to follow them? Shouldn't science teaching standards actually match the rigors of universal scientific progress? Should self-regulation in school curricula break apart from proper science education in the name of self-regulation?"

Massachusetts just got an A- on The State of State Science Standards, a 2012 report released by the Thomas B. Fordham Institute. Although "The Bay State" ranked fourth in the nation, after California (A), District of Columbia (A) and Indiana (A-), the other New England states did poorly: Connecticut ranked 14th, Vermont 24th, Maine 30th, New Hampshire 31st and Rhode Island 33rd.

"*A majority of states' standards remain mediocre to awful*," the report says. "*The average grade across all states is —once again— a thoroughly undistinguished C. In fact, it's a low C*," highlights the Foreword to a 217-page state-by-state scrutiny of the science education expectations in physics, earth and space, and life sciences.

When contrasting the overall performance between 2005 and 2012, a reality emerges: only 11 states improved their grades, usually from F to higher —but still embarrassing— scores, except for the District of Columbia which

---

* Essay first published in *The Standard Times* of New Bedford, March 21, 2012.

went from a C to an exemplary A; 19 states worsen from a B or A to lower grades; and 20 states remained unchanged, half of them around F or D.

In New England, Massachusetts was "degraded" from A to A-, Connecticut and Vermont remained in C, and Maine in D, Rhode Island decreased from C to D, and New Hampshire improved from F to D.

The letter grades corresponded to numeric scores over 10 points; seven for content and rigor of state science standards and three for clarity and specificity. However, the assessment by the Fordham Institute was not about actual performance of students or teachers in physics, earth and space, or life sciences, but exclusively about the science expectations that schools are supposed to meet. Note that the U.S. world placement in math (25th), science (17th) and reading (14th) has been documented in previous studies.

A few questions emerge out of the Fordham results: If schools and teachers stick to a "C average" state of science standards, what quality of education are we really offering? If state science standards guide science education, how do we escape from this loop of poor standards and the expectation to follow them? Shouldn't science teaching standards actually match the rigors of universal scientific progress? Should self-regulation in school curricula break apart from proper science education in the name of self-regulation?

First, the undermining of evolution, a battle not only traceable to the Scopes Trial of 1925, in Tennessee, when the Butler Act (see link to original document, below) declared "*unlawful to teach any theory, in public schools, that denies the Story of the Divine Creation of man as taught in the Bible*," but currently to the 2012 New Hampshire bill proposals to "*require evolution to be taught as a theory, including the theorists' political and ideological view points and their position on the concept of atheism*" or "*require science teachers to instruct pupils that proper scientific inquire results from not committing to any one theory or hypothesis, no matter how firmly it appears to be established*" (links to original documents, below). Both bills are hymns to ignorance.

Second, vagueness in the statements of expectations about what students should actually learn or be able to do after learning; 29 states scored 1 or 0 in the clarity and specificity score of the assessment, out of three points.

Third, poor integration of scientific inquiry, a pernicious malady nationwide; rather than helping students to acquire scientific content through discovery, there is "too much attention to engineering and technology, as well as to "science process skills," which leads to a technical mind-set where true scientific thinking is lacking.

And fourth, poor foundations in mathematical skills; in essence, the most significant tool for modern scientific explorations, that is mathematics, is avoided as the centerpiece of proper science education: a fear of equations and rejection of complexity, or, as one of my excellent students ironically puts it, "there are too many numbers in math."

An illustration by Sarah Samaroo is the only humorous, purposely macabre, aspect of the Fordham report. She depicts on the cover a *Tyrannosaurus rex* grossly salivating and crushing, before gulping, massive paper balls —as if they were carcasses of cellulose— of the "Science Standards." In the background, a mega eruption and in the foreground an asteroid colliding with Earth warns us of the imminent extinction of magnificent science standards that we could have fully possessed or preserved if proper scientific inquiry had been in our minds.

Suggested Readings and Resources

Anti-evolution bills in New Hampshire. (2012). http://ncse.com/news/new-hampshire.

Butler Act. House Bill 185. (1925). http://www.tennessee.gov/tsla/exhibits/scopes/images/Butler%20Act.pdf.

Gross, P. R., Goodenough, U., Haack, S., Lerner, L. S., Schwartz, M. & Schwartz, R. (2005). *The state of state science standards*. Washington DC: Thomas B. Fordham Institute.

Sahlberg, P. (2013). *Finnish Lessons: What Can the World Learn from Educational Change in Finland?* New York, NY: Teachers College Press.

Tennessee vs. John Scopes, "The Monkey Trial." http://law2.umkc.edu/faculty/projects/ftrials/scopes/scopes.htm.

Thomas B. Fordham Institute. (2012). The State of State Science Standards. http://www.edexcellence.net/publications/the-state-of-state-science-standards-2012.html.

Essay 19

AMERICANS WANT CANDIDATES TO DEBATE SCIENCE*

"…Statistics –reliable tools in the scientific method— strongly suggest that Americans want a presidential and congressional debate on science, innovation, health, and the environment, and that such dialog should exclude the personal opinions and beliefs of the candidates. Imagine, at last, a conversation over reality, facts, evidence, and rationality. If science becomes the backbone –better the brain— of candidates, and the voters are literate enough to assess it, a single debate shall suffice to unmask it all..."

85% of Americans want a presidential science debate, although more registered democrats (89%) than republicans (83%) would like a match between President Barack Obama and former Massachusetts Governor Mitt Romney over science-based challenges in healthcare, climate change, energy, education, innovation and the economy. In fact, 84% of likely voters rank science, innovation and healthcare as the third most important topic in a debate, after the economy and taxes, and foreign policy and national security.

These views stretch beyond the presidential elections, with 81% of probable voters also expecting congressional science debates. But what surprised me most about these figures, released by Science Debate Dot Org, was that 81% of the 1000 surveyed adults thought that public policies should be based on science, not the personal opinions or beliefs of elected officials. Really? Does the public prefer presidents and congressional representatives to rationally condition their outlooks to what science says? I love it, because the

* Essay first posted on *Evolution Literacy*, November 5, 2012.

data implies that we can safely approach politicians and the public with facts, and expect broad appreciation for the truth, the backbone of science. Right?

Evolution, climate change, the importance of stem cell research, the benefits of vaccines to public health, the cleanness of clean energy, the dangers of pollution, are all scientific realities –not to mention the impending collision of an asteroid with Earth. But what politicians or the citizens believe about "reality" contradicts the enthusiastic 81% support for an honest conversation about facts.

According to Gallup Poll, 40% of Americans accept evolution. Among them, 60% of democrats or independents versus 30% of republicans think evolution is true. Yet, there is no doubt among scientists that cosmic transformations and Darwinian evolution are factual. Gallup also reports that 58% of the general public think that climate change is occurring, versus 75% of democrats, 53% of independents, and 43% of republicans. But thousands of world researchers, advisors to the Intergovernmental Panel on Climate Change, have compiled chronological evidence of concerning fluctuations in climate.

Evolution aside, which formulation as a scientific reality dates back to, at least, 150 years, and to Charles Darwin's seminal contributions (*On the Origin of Species* 1859 and *The Descent of Man* 1871), and whose on-and-off opponents always lose in the court of science and, frequently, in the court of law, the recent human-made corollaries of climate change are also factual despite the opinions of candidates, ideologies or congregations. But here is relevant trivia reported by Yale University and George Manson University Project/Center for Climate Change Communication: Americans trust President Obama (47%) more than former Governor Romney (21%) as "a source of information about global warming."

Remember that four in every five responders to the Science Debate Dot Org poll preferred science-inspired public policies rather than belief-based decisions. However what is expected from politicians by the public differs from what is observed in the electorate voting behavior. Who is telling the truth about evolution or climate change, President Obama or former Governor Romney? The answer is scientists! And that should be the point of reference for those seeking genuineness: learn what science says about reality, expect politicians to understand and match that view, and cast votes accordingly.

Harris Interactive has surveyed that, despite medical researchers' need of experimentation with stem cells to develop treatment or to prevent diabetes, Alzheimer or Parkinson disease, only 72% of the American public thinks such research should be allowed, in contrast to 82% of democrats, 73% of independents, and 58% of republicans. But misinformation about health can be even more scandalous, for example, one in every five adults believes that vaccines cause autism.

The Pew Research Center has reported 52% of public support to developing alternative sources of energy —to oil, coal and gas; 65% of democrats, 55% of independents, and 36% of republicans agree with this view; not surprisingly 81% of progressives versus 52% of conservatives think that more federal funding should sponsor alternative energy research. And three quarters of the electorate trusts more the Environmental Protection Agency — to research, monitor, set standards and reinforce policies concerning pollution— than the US Congress.

Statistics —reliable tools in the scientific method— strongly suggest that Americans want a presidential and congressional debate on science, innovation, health, and the environment, and that such dialog should exclude the personal opinions and beliefs of the candidates. Imagine, at last, a conversation over reality, facts, evidence, and rationality. If science becomes the backbone —better the brain— of candidates, and the voters are literate enough to assess it, a single debate shall suffice to unmask it all.

SUGGESTED READINGS AND RESOURCES

George Mason University. Center for Climate Change Communication. http://www.climatechangecommunication.org/.

Harris Interactive. Embryonic stem cell research receives widespread support from Americans. http://www.harrisinteractive.com/vault/Harris-Interactive-Poll-HealthDay-2010-10.pdf.

Intergovernmental Panel on Climate Change. http://www.ipcc.ch/.

National poll of likely voters' attitudes about the roles of science and faith in the 2012 election. http://www.sciencedebate.org/assets/downloads/ScienceDebate2012Poll.pdf

Science Debate Dot Org. http://www.sciencedebate.org/.

The Gallup Poll. (2007). Majority of republicans doubt theory of evolution. http://www.gallup.com/poll/27847/Majority-Republicans-Doubt-Theory-Evolution.aspx.

The Gallup Poll. (2009). On Darwin's birthday, only 4 in 10 believe in evolution. http://www.gallup.com/poll/114544/Darwin-Birthday-Believe-Evolution.aspx.

The Gallup Pool. (2011). In U.S., global warming views steady despite warm winter. http://www.gallup.com/poll/153608/Global-Warming-Views-Steady-Despite-Warm-Winter.aspx.

The Pew Research Center for The People and The Press. (2011). Partisan divide over alternative energy widens. http://www.people-press.org /2011/11/10/partisan-divide-over-alternative-energy-widens/?src=prc-headline.

Yale Project on Climate Change Communication. http://environment.yale.edu/ climate-communication/.

Essay 20

Darwin Day Awaits Designation by the US Congress*

> "...Darwin Day... signifies the celebration of the achievements of reason, science, and the advancement of human knowledge. The igniting moments in human history when light was brought into our own origins, when understanding that ordinary apes, like Homo, were capable of the extraordinary, of discovering the truth and debunking obscurantism; yet we still struggle to make science the sole guiding star in our survival decisions, the reliable source of concern and joy, the toolkit to plan our departure from Earth –before our Sun in agonizing heat engulfs its nearest orbiting planets– and seek home somewhere else in the cosmos."

On January 22, 2013, representative Rush Holt (D) of New Jersey, and seven of his colleagues, introduced bill H.Res.41 to Congress expressing support for designation of February 12 as Darwin Day.

The Congressional Record of the US House of Representatives (p. H437), dated February 12, 2013, reports: 2 p.m. Prayer; approval of the Journal of the last day's proceedings; Pledge of Allegiance... Darwin Day, among other items. In one minute, Mr. Holt summarized for the House the purpose of the bill, which was later referred to the House Committee on Science, Space and Technology (more on this Committee below).

Why Darwin Day? The bill "whereases" explain:

> "Charles Darwin theory of evolution by the mechanism of natural selection, together with the monumental amount of scientific evidence he

---

* Essay first posted on *Evolution Literacy*, May 7, 2013.

> compiled to support it, provides humanity with a logical an intellectually compelling explanation for the diversity of life on earth."
>
> "The validity of Darwin's theory of evolution by natural selection is further strongly supported by the modern understanding of the science of genetics."
>
> "It has been the human curiosity and ingenuity exemplified by Darwin that has promoted new scientific discoveries that have helped humanity solve many problems and improve living conditions."
>
> "The advancement of science must be protected from those unconcerned with the adverse impacts of global warming and climate change."
>
> "The teaching of creationism in some public schools compromises the scientific and academic integrity of the United States education systems."
>
> "Charles Darwin is a worthy symbol of scientific advancement on which to focus and around which to build a global celebration of science and humanity intended to promote a common bond among all of Earth's peoples."
>
> "February 12, 2013, is the anniversary of the birth of Charles Darwin in 1809 and would be an appropriate date to designate as Darwin Day: Now, therefore, be it Resolved, that the House of Representatives (1) supports the designation of Darwin Day, and (2) recognizes Charles Darwin as a worthy symbol on which to celebrate the achievements of reason, science, and the advancement of human knowledge."

Bill H.Res.41, itself, embodies the never-ending battle against irrationalism, the latter vividly present in the views of those who see evil in truth and menace in the realities discovered by science. Take, for example, last year's remarks by congressman Paul Broun (R), from Georgia, a physician and member of the House of Representatives Committee on Science, Space and Technology (the very Committee to which the "Darwin Day bill" was referred), who declared: "*God's word is true. I've come to understand that. All that stuff I was taught about evolution and embryology and the big bang theory, all that is lies straight from the pit of Hell.*"

And Mr. Broun went on, as documented in video watched worldwide (see link to video below): *"It's lies to try to keep me and all the folks who were taught that from understanding that they need a savior." "You see, there are a lot of scientific data that I've found out as a scientist that actually show that this is really a young Earth." "I don't believe that the earth's but about 9,000 years old." "I believe it was created in six day as we know them." "That's what the Bible says."*

But what shocked me most about Mr. Broun's remarks at the 2012 Sportsman's Banquet at Liberty Baptist Church in Hartwell, Georgia, was not the interactive mood of the audience excitedly identifying with and encouraging the speech to ascend to pulpit climax, but how startled looked the dozens of antlered-and-dead deer-trophies casted on the background which, unlike "deer in the headlights" helpless to the imminent collision with the unseen, seemed almost responsive and eager to stampede.

Ironically, the quadrupeds themselves symbolize evolution by natural and sexual selection, magnificent concepts explained to the Victorian public by Darwin on *The Origin of Species* (1859) and *The Descent of Man* (1871).

Darwin reasoned about antler evolution: *"...With stags of many kinds the branching of the horns offers a curious case of difficulty; for certainly a single straight point would inflict a much more serious wound than several diverging points." "The suspicion has therefore crossed my mind that they may serve partly as ornaments. That the branched antlers of stags, as well as the elegant lyrated horns of certain antelopes, with their graceful double curvature, are ornamental in our eyes, no one will dispute. If, then, the horns, like the splendid accoutrements of the knights of old, add to the noble appearance of stags and antelopes, they may have been partly modified for this purpose, though mainly for actual service in battle.*"

Indeed, Darwin Day —now awaiting action by the House Committee on Science, Space and Technology— signifies the celebration of the achievements of reason, science, and the advancement of human knowledge. The igniting moments in human history when light was brought into our own origins, when understanding that ordinary apes, like *Homo*, were capable of the extraordinary, of discovering the truth and debunking obscurantism; yet we still struggle to make science the sole guiding star in our survival decisions, the reliable source of concern and joy, the toolkit to plan our departure from Earth —before our Sun in agonizing heat engulfs its nearest orbiting planets— and seek home somewhere else in the cosmos.

SUGGESTED READINGS AND RESOURCES

Congressional Record of the US House of Representatives (p. H437). (2013). http://beta.congress.gov/crec/2013/02/12/CREC-2013-02-12-pt1-PgH437-7.pdf.

Darwin Online, The world's largest and most widely used resource on Charles Darwin. http://darwin-online.org.uk/.

House Committee on Science, Space and Technology. http://science.house.gov/.

Representative Paul Broun (R) video on evolution, embryology and big bang theory. http://www.youtube.com/watch?v=ZBy3MbP4WDo.

Essay 21

CAN ATHEISTS BE OUR LEADERS?*

> "…It is time for our modern societies to accept the open participation of atheists and agnostics in building our democracies, more so if —as demonstrated by [the] polls— they are among the most educated citizens... The human experience consists in building equality for all and… in hearing the voice and benefiting from the talent of the nonbelievers."

Atheists and agnostics are among the most educated citizens in the United States. They rank highest not only in knowledge about science, American history, literature, politics and the role of religion in public life, but also in awareness about world religions.

Using a 32-question survey on religious knowledge, the Pew Forum on Religion and Public Life, a non-governmental organization based in Washington D.C., interviewed 3,400 adults during May and June of 2010. A 78-page report of the significant findings was released this past September 28, 2010.

Knowledge about religion correlated with level of education. Responders holding a post-graduate degree knew, on average, 22 out of the 32 questions in the survey; college graduates responded correctly to 20 questions; those attending college were right about 17 to 19 questions; and the high-school-educated —or less— were correct in only 10 to 12 questions.

Atheists and agnostics followed by Jews and Mormons ranked consistently higher in the overall assessment of their religious literacy than evangelical Protestants and Catholics.

---

* Essay first published in *The Standard Times* of New Bedford, November 6, 2010.

If atheists and agnostics are highly educated, would Americans elect them as their leaders? Apparently not; according to Gallup Poll (2007, see link below), atheists rank last with only 45% voters to favor them in a potential presidential election, followed by homosexuals, who would theoretically receive 55% of support, or candidates of "72 years of age" (57%), or who are "married for the third time" (67%), or Mormon (72%), Hispanic (87%), a woman (88%), Jewish (92%), black (94%, it already happened), or Catholic (95%).

Sixty-seven percent of liberals would vote for a qualified atheist if he or she runs for president of the United States, but only 29% of conservatives would do it. Indeed, political ideology determines voting preferences for a "non-traditional candidate," and atheists rank last regardless of being among the most literate Americans. Moreover, liberal, moderate and conservative voters would prefer any other type of candidate over an atheist (Gallup Poll, 2007).

Twenty million Americans, or 7%, are either certain that God does not exist (atheists), or are not sure about it (agnostics), which contrasts with the high levels of religiosity among most of the general population. Seventy-three percent are convinced of the existence of a deity, 14% think that God probably exists and have little doubt about it, and 5% believe in God but have a lot of doubt about it (Gallup Poll, 2006). Religiosity, however, decreases with educational attainment; highly educated people are less religious than the least educated.

According to the Pew Global Attitudes Project (2007), 57% of Americans think that God is necessary for morality, yet there is no indication that atheists and agnostics are less moral than the general or religious populations.

In fact, the list of world-prestigious American intellectuals who have admitted to be atheists or agnostics is impressive. Here are some from the 19th and 20th centuries: Ralph Waldo Emerson (author and poet), Henry David Thoreau (philosopher), Andrew Carnegie (philanthropist), Mark Twain (author), Pearl S. Buck (author), Thomas Edison (inventor), Clarence Darrow (lawyer), Carl Van Doren (English professor and biographer of Benjamin Franklin), and Ernest Hemingway (novelist).

The list above does not include the 93% of the current and prominent members of the U.S. National Academy of Sciences (Larson and Witham, 1998), neither the 70% of the 630,000 faculty teaching full-time at colleges and universities in the United States who admit to be nonbelievers (Ecklund and Scheitle, 2007; Gross and Simmons, 2009).

Like any highly educated citizens, atheists and agnostics are probably concerned about illiteracy trends in the United States, and not only regarding world religions knowledge —where they ranked highest above all believers— but also international trends on the ranking of our youth in mathematics, reading and science, where the United States placed 26th, 15th and 21st among 57 other nations, respectively (data from the Organisation for Economic Co-operation and Development, reports 2003/9, links below).

And we can forecast that atheists and agnostics, who rank top in understanding the legal separation between church and state (82%, according to the Pew survey), would support the teaching of evolution in public schools and oppose the smuggling of intelligent-design creationism into the education curriculum.

It is time for our modern societies to accept the open participation of atheists and agnostics in building our democracies, more so if —as demonstrated by national polls— they are among the most educated citizens. Their lack of religious affiliation and identity should not discourage them from contributing to significant public service as secular humanists, nor should it deter the public from electing them.

The human experience consists in building equality for all and, in this particular case, in hearing the voice and benefiting from the talent of the nonbelievers.

SUGGESTED READINGS AND RESOURCES

Ecklund, E. H. & Scheitle, C. P. (2007). Religion among academic scientists: distinctions, disciplines, and demographics. *Social Problems*, 54, 289-307.

Gross, N. & Simmons, S. (2009). The religiosity of American college and university professors. *Sociology of Religion*, 70, 101-129.

Larson, E. J. & Witham, L. (1998). Leading scientists still reject God. *Nature*, 394, 313.

Organisation for Economic Co-operation and Development. http://www.oecd.org/index.htm.

Paz-y-Miño-C, G. & Espinosa A. (2012). Atheists' knowledge about science and evolution. *Secular World*, 18, 33-36.

The Gallup Poll. (2006). Who believes in God and who doesn't? http://www.gallup.com/poll/23470/Who-Believes-God-Who-Doesnt.aspx.

The Gallup Poll. (2007). Some Americans reluctant to vote for mormon, 72-year-old presidential candidates. http://www.gallup.com/poll/ 26611

/Some-Americans-Reluctant-Vote-Mormon-72YearOld-Presidential-Candidates.aspx.

The Pew Global Attitudes Project. (2007). http://pewglobal.org/files/pdf/258.pdf.

The Pew Research Center for The People and The Press. Religion and Public Life Project. (2011). How much do you know about religion? http://features.pewforum.org/quiz/us-religious-knowledge/.

The Pew Research Center. The Pew Forum on Religion and Public Life. (2012). "Nones" on the rise: one-in-five adults have no religious affiliation. Washington, DC: The Pew Reseach Center. http://www.pewforum.org/Unaffiliated/nones-on-the-rise.aspx.

The Pew Research Center. The Pew Forum on Religion and Public Life. (2012). The global religious landscape: a report on the size and distribution of the world's major religious groups as of 2010.

WIN-Gallup International. (2012). Global index of religiosity and atheism. www.Gallup-international.com.

EPILOGUE

> "…we still struggle to make science the sole guiding star in our survival decisions, the reliable source of concern and joy, the toolkit to plan our departure from Earth —before our Sun in agonizing heat engulfs its nearest orbiting planets— and seek home somewhere else in the cosmos."

Millennia after battling their obstinate rivals —superstition and faith— science and its major discovery ever, that of evolution, stand upright, as if they too had transformed gradually from earlier forms, like those organisms that left the ocean, ventured into land and later walked away, bipedally, from their ancestral arboreal habits. Yes, science and evolution strong, by themselves, with no need of sponsorship by a proximate or distant "Architect, Creator or Intelligent Designer," or by in-principle or in-practice creationism. No "belief" endorsement is necessary for science/evolution's persistence in society, except for the careful substantiation of new empirical observations, re-assessment of the old discoveries, and endless re-testing of hypotheses by skeptical, accredited researchers.

"*Evolution is the language of the cosmos*," as my research collaborator, Avelina Espinosa, and I wrote in "*On The Theory of Evolution vs. The Concept of Evolution*" (Evolution, Education and Outreach 2010), while arguing with colleagues about the application of the "concept of evolution" to the universe; a not-new metaphor, or proposal, since astrophysicists have discussed "cosmic evolution" for decades.

Once embraced by all, this truly universal language —scientific rationalism/empiricism and evolution— shall lead us to a more cohesive understanding of nature and of our amazingly diverse human condition. Humanity's ultimate challenge will be to collectively embrace reality, with no stoppers of thought or restrains on logic.

INDEX

A

B

C

D

E

F

G

H

I

Q

R

S

U

V

W

Y